Beyond the
Theory of Constraints

Beyond the Theory of Constraints

How to Eliminate Variation and Maximize Capacity

by William A. Levinson

New York

Copyright © 2007 by Productivity Press, a division of The Kraus Organization Limited

All rights reserved. No part of this book may be reproduced or utilized in any form or by any means, electronic or mechanical, including photocopying, recording, or by any information storage and retrieval system, without permission in writing from the publisher.

Most Productivity Press books are available at quantity discounts when purchased in bulk. For more information, contact our Customer Service Department (888-319-5852). Address all other inquiries to:

Productivity Press
444 Park Avenue South, 7th Floor
New York, NY 10016
United States of America
Telephone: 212-686-5900
Fax: 212-686-5411
E-mail: info@productivitypress.com
ProductivityPress.com

Library of Congress Cataloging-in-Publication Data

Levinson, William A., 1957–
   Beyond the theory of constraints : how to eliminate variation and maximize capacity / by William A. Levinson.
     p. cm.
Includes bibliographical references and index.
ISBN 978-1-56327-370-4 (alk. paper)
   1. Production control. 2. Process control. 3. Theory of constraints (Management) I. Title.
TS155.L3668   2007
658.5—dc22

2007022643

11 10 09 08 07   5 4 3 2 1

CONTENTS

Preface	ix
ONE The Theory of Constraints	1
Implications: Managerial Economics	2
Opportunity Costs	3
The Economic Impact of Stoppages in the Constraint	5
Performance Measurements	6
Dysfunctional Performance Measurements and their Effects	7
The Equipment Utilization Metric	11
Sunk Costs	12
The Return on Investment (ROI) Metric	13
Variable Costs in the Theory of Constraints	14
Finding the Constraint	17
Linear Programming	19
Simplex Method	20
The Market Constraint	27
Contractual Requirements and Overtime	28
Summary: The Theory of Constraints	29
TWO Production Control: Pull versus Pig-Swallowing	33
Just-in-Time Production Control	34
Synchronized Production and Takt Time	35
Kanban Systems	39
Drum-Buffer-Rope	42
Buffer Management	43
Simplified Market Pull	45
CONWIP	46
Statistical Throughput Control	47
Flow: Keep the Work Moving	49
Flow Lines and Paced Lines	50
The Plug Flow Reactor: A Model for the Ideal Situation	50
The Flying Boxes: Flow at the Start of the 20th Century	52
Summary: Pull Production Control	54

CONTENTS

THREE Variation   57

Variation Sources   57
    "Waiting to Match" and Supply-Chain Dependence   59
    Batching   60
    Quality and Reliability Problems   61
Effects of Variation: The Matchsticks-and-Dice Exercise   62
Effect of Variance: Performance Characteristics of Single-Server Queue   67
Summary: Variation   70

FOUR Variation Reduction   73

Subdivision of Labor   73
Value Stream Analysis   77
    Workflow Analysis   78
    Cycle Time Accounting   80
    "Five Whys" Technique   82
Single-Unit Processing   84
    Process Batching and Transfer Batching   84
    Batching Increases Cycle Time and Variation   85
    Parallel Process Batch Processes Are Harder to Control   87
    Alternatives to Batch Processing   88
Single-Minute Exchange of Die   90
    Military Origins of SMED   92
    Split-Thread Bolts and Quick-Clamping Flanges   95
Work Cells and Unitary Machines   97
    Spaghetti Diagram   97
    Work Cells   98
    The Unitary Machine   100
    Dedicated Equipment by Product Family   103
Summary: Variation Reduction   103

CONTENTS

FIVE Productivity Improvement		107
Friction		107
False Economy		109
False Economy of Cheap Equipment		110
False Economy of Cheap Labor		111
False Economy of Cheap Purchases		114
False Economy of Cheap Working Conditions		115
False Economy of Cheap Procedures		116
Motion Efficiency		118
In-line Quality Control		120
Error-Proofing (Poka-Yoke)		122
Self-Check Systems		122
Source Inspections		124
Autonomation (Jidoka)		125
Summarizing In-line Quality Control		126
Preventive Maintenance		126
Supply Chain Management		128
The Beer Game		128
Sources of Instability in Supply Chains		129
Logistics		131
Summary: Productivity Improvement		132
Appendix: Linear Programming in Microsoft Excel		135
Bibliography		143
Index		147
About the Author		155

PREFACE

Eliyahu Goldratt and Jeff Cox's *The Goal* asks, "Why do you think it is that nobody, after all this time and effort, has ever succeeded in running a balanced plant?"[1] Henry Ford claims to have done so: "The idea is that a man must not be hurried in his work—he must have every second necessary but not a single unnecessary second."[2] Ford's apparent success in doing what *The Goal* shows to be impossible prompted this book's development.

When manufacturing engineers think of variation, critical-to-quality product characteristics are the first things that come to mind. The concept of variation in product characteristics is, in fact, central to the quality sciences. This variation is not, however, the variation that prevents operation of a balanced factory at close to 100 percent capacity. Variation in processing times and material transfer times either wastes capacity or requires large inventories as insurance against its effects. It is, therefore, necessary to state the following proposition at the outset:

- Variation in product characteristics causes rework and scrap. This is the familiar random or common-cause variation whose effects are shown by measurement histograms. The process standard deviation is the basis of the control limits for statistical process control charts.
- Variation in processing and material transfer times is the root cause of longer cycle times, higher inventories, and an inability to run a balanced factory at close to 100 percent capacity.

The matchsticks-and-dice simulation in *The Goal* illustrates the latter variation's effects. The simulation also shows Ford's proposition to be an obvious formula for a deranged nightmare in which inventory overruns the factory while cycle time in queue becomes infinite. As utilization approaches 100 percent, cycle time in queue (and, therefore, inventory) will indeed approach infinity—unless variation in processing times and material transfer times approaches zero. Ford's production system was designed explicitly to suppress this kind of variation, and his success demands close investigation of the methods he used. Furthermore, Toyota's *heijunka* (level scheduling and production smoothing) concept reflects both the need and the ability to suppress the "random" variation suggested by *The Goal*'s matchsticks-and-dice factory simulation.

"JIT (just-in-time) is also helpless unless downstream production steps practice level scheduling (*heijunka* in Toyota-speak) to smooth out the perturbations in day-to-day order flow unrelated to actual customer demand. Otherwise, bottlenecks will quickly emerge upstream and buffers ('safety stocks') will be introduced everywhere to prevent them."[3]

It is this author's conclusion that *The Goal*'s matchsticks-and-dice exercise is an excellent device for teaching the effects of variation on throughput and inventory. The example may also teach the unintended lesson that the factory is at the mercy of this variation. A die roll suggests unavoidable *random variation* (also known as *common cause variation*), but the "workstation" is nonetheless *capable* of processing six units. This book's purpose is to teach the reader how to identify and remove the variation, and thereby roll a six every time.

The book is organized as follows:

- Chapter 1 is an overview of the Theory of Constraints (TOC) and also covers the engineering and managerial economic aspects of TOC.
- Chapter 2 covers pull production control methods such as kanban and synchronous flow manufacturing's drum-buffer-rope (DBR) system.
- Chapter 3 illustrates the effect of variation in processing and material transfer times, and shows why this variation prevents achievement of 100 percent utilization.
- Chapter 4 describes methods for reducing variation in processing and material transfer times. Some of the material presented here overlaps with the theme of Chapter 5 because techniques that suppress variation often improve productivity and vice versa.
- Chapter 5 discusses methods for increasing productivity and reducing cycle time. These methods are useful for elevating the constraint (increasing its capacity) and may also reduce variation.

THE MATCHSTICKS-AND-DICE SIMULATOR.

As an added feature, visit http://www.ct-yankee.com/lean/toc_dice.html for a downloadable program for simulating the matchsticks-and-dice exercise from the classic Goldratt and Cox narrative *The Goal* (North River Press,

1984). The program was compiled with Visual Basic Service Pack 6, and should work in most Windows operating systems. Installation requires the cabinet file (Dice.CAB), Setup.exe, and the list file SETUP.LST, all of which must be placed in the same directory.

The roll of a single die is simulated for each workstation. The workstation can process the lesser of the die roll and the number of pieces in queue. Each workstation starts with four pieces in stock to reduce the time necessary for the system to reach a supposedly steady state.

The user can select anywhere between one and six workstations, and the simulation interval in seconds. Long intervals allow viewers to see the details for each turn of the simulation, while short ones allow them to see what happens after several thousand turns. A bar chart displays the inventory at each station (and possibly large inventory "bubbles," as have been seen in real factories). The neighboring graph shows the average throughput and the total inventory in the system.

ENDNOTES

1. Goldratt and Cox, *The Goal*, 1992, p. 86
2. Ford, *My Life and Work*, p. 82
3. Womack and Jones, *Lean Thinking*, p. 58

ONE

The Theory of Constraints

The Theory of Constraints (TOC) states that the capacity of any manufacturing process is limited by its capacity-constraining resource (CCR) or constraint. *The Goal* uses a Boy Scout hike to illustrate the principle that no set of interdependent activities can operate more quickly than its slowest part (Figure 1-1).

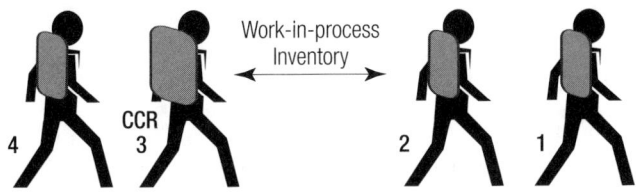

Figure 1-1. A line of hikers illustrates the Theory of Constraints.

In this analogy, the total distance traveled symbolizes the throughput or total production, while the length of the column symbolizes inventory. Each man symbolizes a production operation, and all four operations must be completed before throughput occurs. The third hiker has a bigger load, so he is slower than the rest. He is the constraint or CCR.

The first hiker symbolizes production starts. The distance between the second and third (inventory) will continue to increase if the first and second hikers walk at their maximum pace. *The Goal* uses the analogy of a rope between the third and first hiker to describe a pull-type production control system in which the CCR sets the pace for production starts. This shortens the column's length (work-in-process) because the leader will not walk any faster than the constraint. Some distance is allowed between the slowest hiker and the man in front of him. If the slowest hiker has to pause or stop because the man ahead of him varies his pace slightly or stumbles (for example, an equipment stoppage), he can never make up the lost distance. In synchronous flow manufacturing, this distance is known as the *buffer*.

A group of hikers cannot arrive at its destination more quickly than its slowest member, and a manufacturing process cannot work more quickly than its slowest operation. This seems obvious, but it has profound implications for managerial economics and business decision making, as the following section discusses.

IMPLICATIONS: MANAGERIAL ECONOMICS

> *Money is dear; human life is still dearer; but time is dearest of all. One minute decides the outcome of a battle, one hour the success of a campaign, one day the fate of empires. . . . I operate not by hours but by minutes.*
> —ALEKSANDR V. SUVOROV[1]

Basic principle: SPEED KILLS (competitors).

Perhaps one of the most important lessons to take from this chapter is that *time is a cost*. Even the best forms of managerial economic analysis—that is, the ones that consider only marginal or differential cash flows (a concept that this chapter discusses in some depth)—can overlook this.

The old adage that "time is money" understates the case. Time is money not only because workers must be paid for their time, but also because time can convey a decisive competitive advantage in the marketplace. Businesses often focus on quantifying the time it takes to do a job because workers are paid by the hour. An Internet search finds many entries for "time accounting," but the Web pages' primary concern is with charging labor or professional time to different projects and activities: that is, time expended by people. This book will later consider a form of time accounting that addresses time expended by parts.

Most business leaders would become very agitated if they discovered that their people stood around doing nothing for 95 percent of the workday, but those same people are completely indifferent to parts that spend 95 percent of their time doing nothing. No one pays much attention to parts in a storeroom or warehouse because no worker's time can be charged to them. Long lead times or an inability to make just-in-time deliveries, however, can result in business going to a competitor. The underlying problem is, of course, failure to use cycle time as a performance measurement.

Failure to treat cycle time as a cost can also result in false economies. Sending a truck somewhere with less than a full load increases marginal costs, because the driver's time and the truck's fuel cost real money. On the other hand, waiting for a full load can increase lead times while aggravating the effects of batching and queuing. Only the recognition of time as a cost causes people to think about truck sharing and other methods for moving partial loads economically.

Time may indeed be the dearest of all commodities, and time lost at the constraint is lost forever. The CCR (by definition) lacks reserve capacity with which to make up any shortfall. This lack of reserve capacity has some very interesting and important effects on the process's managerial economics. Rework in the constraint can be even more costly than scrap prior to the constraint because the CCR has no time to spare. Scrap subsequent to the constraint is equally disastrous because the constraint has no spare capacity with which to replace the losses. The concept of opportunity cost is very important in managerial economics, and the following section explains it in detail.

Opportunity Costs

Benjamin Franklin, perhaps the true intellectual father of lean manufacturing, wrote in *The Way to Wealth*, "Lost time is never found again. . . ."[2] Rephrased in Theory of Constraints language, this adage becomes "time lost at the constraint is lost forever." In *Poor Richard's Almanack* Franklin observed, "He that idly loses 5s. [shillings] worth of time, loses 5s., and might as prudently throw 5s. into the river."[3] This underscores exactly what happens when the CCR because of lack of work (starving the constraint), lack of personnel, or a mechanical stoppage. Henry Ford, who cited Franklin as an influence on his own thought process, laid out the principle as follows:

Time waste differs from material waste in that there can be no salvage. The easiest of all wastes, and the hardest to correct, is this waste of time, because wasted time does not litter the floor like wasted material.[4]

This statement further underscores the desirability of time accounting, not for people, but for work-in-process (WIP) and cycle time. Meanwhile, opportunity cost relates almost exclusively to the loss of time as opposed to loss or waste of materials:

> ... one of the most fundamental distinctions between two general concepts of cost is that between accounting (absolute) cost and opportunity (alternative) cost.... Opportunity cost is concerned with the cost of forgoing certain opportunities or alternatives in favor of pursuing others.[5]

In simple terms, this statement posits that money invested in one enterprise cannot be invested in another. When computing the cost of homeownership, many people consider only their monthly mortgage payment, taxes, homeowner's insurance, and maintenance outlays. What they generally do not consider is that money invested in a home cannot be invested in the stock market or put in the bank, so the true cost of owning the house must include whatever interest or dividend could have been earned on the capital outlay. On the other hand, not having to pay rent can be viewed as after-tax income.

James L. Rigg extends the concept of opportunity cost to income foregone because an item is not in stock and discusses, in particular, perishable goods or those subject to obsolescence.[6] Fruits and vegetables not sold within a specific time limit spoil, and yesterday's newspaper is worthless. On the other hand, not having the item in stock when the customer wants it means a lost sale and lost profit. These examples illustrate the importance of accurate demand forecasting when a product is perishable; they also show why make-to-order capability can provide an overwhelming competitive advantage. If, for example, it were practical to print newspapers to order at the point of sale, there would be no risk of stockouts or unsold inventory.

In many cases, the capacity is fixed and must accommodate anticipated demand. Consider, for example, a passenger aircraft with 100 seats or a hotel with 100 rooms. If the airline sells only 80 seats for a given flight or the hotel gets only 80 guests on a given day, the unused 20 units of capacity will *never* be sold or used. The opportunity to profit from those seats or rooms is lost forever, just as unused or wasted constraint capacity is lost forever. When time is wasted at a manufacturing or service constraint, the lost opportunity to make and sell another unit can never be recovered.

The Economic Impact of Stoppages in the Constraint

The true impact of any stoppage in the constraint, whether from lack of work (known as starving the constraint) or rework, can far exceed the accounting cost. The best way to understand this is to examine cash flows. Assume that the product sells for $110, and that labor, material, and overhead costs are as shown in Table 1-1.

Table 1-1. Labor, material, and overhead costs.

	OPERATION		
	1	2 (constraint)	3
Raw materials	$5	$5	$5
Labor	$5	$5	$5
Overhead	$10	$10	$10

The cost accounting system would charge $20 for a rework at the constraint: $5 each for materials and labor, and $10 for overhead. The accounting system might also report the non-use of materials and labor at the other operations (because the rework causes one less item to be produced) as volume variances. The cash flow perspective of this situation is quite different, as shown in Table 1-2.

Table 1-2. Cash flow cost of a rework in the constraint.

Raw material needed to rework the item	$5
Loss of a sale due to waste of capacity at the constraint	$110
Material not used in operation 3 (because one less unit comes from the constraint)	($5)
Material not used in operation 1 (as one less unit is started to replenish the constraint's inventory buffer)	($5)
Total rework cost (cash flow)	$105

In other words, a rework in the constraint actually costs the organization $105 in real money as opposed to $20 in cost-accounting money. This is an important point. On one hand, labor and overhead are not real money, because the organization is paying for the workers whether they are working or not. The fact that the constraint's operator reworks a nonconforming piece instead of shipping a good one does not actually add to the factory's expenses. The factory also pays for utilities, insurance, and other fixed costs no matter what happens at the constraint. The actual added cost of the rework is $5 for materials, as the finished unit will have consumed $20 instead of $15. Furthermore, as the factory's throughput is reduced by one, the first and third operations will each use $5 less in materials. Factories, however, do not make money by reducing their throughput, and the *differential cash flow* because of this rework is a $105 loss.

A more detailed discussion of differential costs and cash flow is presented later in this chapter. The following section addresses the Theory of Constraints' simple but sensible performance measurements.

PERFORMANCE MEASUREMENTS

> *"The first man had his three wishes, yes,"* was the reply. *"I don't know what the first two were, but the third was for death. That's how I got the paw."*
> —W. W. Jacobs, "The Monkey's Paw"

"Be careful what you wish; you might get it" is an ancient proverb that many legends and folk tales illustrate. Two basic principles apply to all business performance measurements.

- Performance measurements drive behavior. Management *always* gets what it asks, but what it asks is not always what it wants. This is the lesson of W. W. Jacobs' horror story "The Monkey's Paw," in which a sinister talisman finds the most malicious possible ways to grant its owner's wishes. Unlike the talisman, real-world employees are seldom malevolent, but they do behave rationally in response to what has been requested and will be measured. Generally, this means behaving in a manner that brings rewards. Dysfunctional performance measurements evoke dysfunctional and destructive behavior.

- Metrics should be *objective measurements* of business process performance that are easily understood by all participants. They should also be few in number.

The Theory of Constraints uses only three simple metrics: throughput, inventory, and operating expense.

- Throughput consists of finished goods with customers, as opposed to finished goods that are warehoused in the hope of finding customers. More is better.
    - Taiichi Ohno added the measurement of cycle time to that of throughput. "All we are doing is looking at the time line from the moment the customer gives us an order to the point when we collect the cash. And we are reducing that time line by removing the non-value-added wastes."[7]
    - Cycle time as a fourth TOC metric is partially redundant with inventory because Little's Law states that cycle time equals inventory divided by throughput. TOC's metrics already encourage improvements in two of these factors, which should automatically optimize the third. Nonetheless, the introduction of cycle time as an explicit metric provides the workforce with an entirely new perspective. Looking for ways to reduce inventory will also lead to reductions in cycle time, while looking for ways to reduce cycle time will similarly reduce inventory.
- Inventory includes all investments in items the factory hopes to sell. Less is better even though traditional cost accounting systems treat inventory as an asset.
- In the case of operating expense, less is also better. This idea was reinforced by Ohno, who noted that passing costs on to customers is not a viable business strategy.

Nowhere do measurements like "equipment utilization," "overhead absorption," or "return on investment" appear in the Theory of Constraints, and for good reason. They are all capable of wreaking as much mayhem on a company as the mythical monkey's paw inflicted on its owners. The following sections provide some examples.

Dysfunctional Performance Measurements and their Effects

"A cost accounting system is not a suicide pact."[8]

Henry Ford created what is now known as *lean manufacturing*, but his contemporary Alfred Sloan may have sown the seeds for the downfall of the U.S. auto industry. James Harrington[9] describes how Sloan and his treasurer, G. Donaldson Brown, created the familiar cost accounting system that values inventory almost as highly as cash, an approach that encourages the production of inventory whether it is needed or not. Furthermore, the Sloan accounting system "allows products to be 'sold' from one internal department to another, so that the sending department makes a profit on the transaction even though no money changes hands. This inflates an organization's value while it increases expenses, scrap and rework costs, and surplus parts costs."

The author adds that Sloan's accounting system has been codified into U.S. tax laws, but further notes that the only time that an organization makes a genuine profit is when a customer buys its product or service. This is, of course, the Theory of Constraints' definition of *throughput*.

The internal sale of work-in-process can have entertaining, if costly, effects. Alan Robinson cites an example in which one department of a food processing plant filled boxes with ingredients. The boxes were then sealed, labeled, and sent to a variety food packaging department less than 100 yards away, where they were torn open and unpacked so the ingredients could be repackaged for retail sale. When an operations management student made the very sensible observation that not packaging the intermediate ingredients in the first place would save money and labor, she was told that this was impossible because the variety pack department had to "buy" the ingredients from the main production line.[10] This is a classic example of the old joke about digging a hole and filling it in (or the one about two people digging side by side and throwing the dirt into each other's holes as shown in Figure 1-2). It becomes far less funny after one reviews some real-world examples.

Charles Standard and Dale Davis cite an engine factory in southeastern Michigan whose production manager is measured against a goal of 1,800 engines a day even though the downstream customer needs only 1,600. The manager fulfills his superior's wishes by making 1,800 engines per day until he has thousands of surplus units. Then the plant shuts down for a couple of weeks until the downstream customer consumes the inventory.[11]

Figure 1-2. Effect of a dysfunctional cost accounting system.

Standard and Davis also cite a purchasing department that is measured on its ability to get low prices on purchased materials. Transportation costs, however, are not charged to the purchasing department but to the assembly plant where the materials are used. The purchasing department might then select a supplier to save a penny per unit of cost while adding five cents per unit to the delivery cost. The managers who set the performance measurement, "Get us the lowest possible price on purchased items," are getting exactly what they *wish* even though the resulting higher overall expenses are not what they *want*.

Dysfunctional purchasing has actually been a problem for more than 200 years. As Benjamin Franklin warned in *The Way to Wealth*:

> You call them goods; but, if you do not take care, they will prove evils to some of you. You expect they will be sold cheap, and, perhaps, they may [be bought] for less than they cost; but, if you have no occasion for them, they must be dear to you. Remember what Poor Richard says, "Buy what thou hast no need of, and ere long thou shalt sell thy necessaries." And again, "At a great penny worth pause a while." He means, that perhaps the cheapest is apparent only, and not real; or the bargain, by straightening thee in thy

business [that is, tightening your circumstances by tying up your cash], may do thee more harm than good. For in another place he says, "Many have been ruined by buying good penny worths."[12]

Henry Ford expressed exactly the same idea in *My Life and Work*:

We have carefully figured, over the years, that buying ahead of requirements does not pay—that the gains on one purchase will be offset by the losses on another, and in the end we have gone to a great deal of trouble without any corresponding benefit. . . . We do not buy less if the price be high and we do not buy more if the price be low. We carefully avoid bargain lots in excess of requirements. It was not easy to reach that decision. But in the end speculation will kill any manufacturer. Give him a couple of good purchases on which he makes money and before long he will be thinking more about making money out of buying and selling than out of his legitimate business, and he will smash. The only way to keep out of trouble is to buy what one needs—no more and no less.[13]

In practice, it may be acceptable to buy bulk quantities of parts or raw materials that are certain to be used. Ford himself stored enormous quantities of iron ore next to his Dearborn steel mills because (1) iron ore does not spoil or become obsolete, and (2) ice could close the Great Lakes to shipping during winter. The latter consideration was doubtlessly the most important. Ford wrote that the cycle time for conversion of iron ore into automobiles was only 33 hours, so a shortfall of coal or iron ore would have shut down the plant in less than two days.[14] As shown by this example, it may be acceptable to violate Franklin's and Ford's general advice on buying only to meet immediate needs if:

- The inventory will not spoil or become obsolete.
- The inventory is not a possible hiding place for quality problems that will be discovered only when the inventory goes into the production line.
- The discount for purchasing large quantities in advance of needs is greater than the inventory's carrying costs.

When deciding whether to purchase or build in advance of needs, first identify the location of what Wallace Hopp and Mark Spearman call the *push-pull interface*.[15] This point is effectively where product differentiation begins. Before it reaches the push-pull interface, WIP is similar to embryonic stem cells that can grow into any body part. Afterward, its flexibility becomes very limited. Hopp and Spearman cite clothing manufacturer Benetton, which produces undyed sweaters to stock and then dyes them to order. There is little risk of the undyed sweaters becoming obsolete, and dyeing to order ensures that none will go unsold because customers don't like the colors.

CafePress uses a similar model. The company keeps an undifferentiated stock of blank garments, coffee mugs, posters, bumper stickers, and so on. It prints designs on these items only when it receives an order. Other online retailers have also adopted this business model.

The Equipment Utilization Metric

"Equipment utilization" is another dangerous and destructive performance measurement, and it comes from the deadly mistake of letting the cost accounting system run the factory. Cost accountants must report income and expenses by methods that are acceptable to the Internal Revenue Service and Securities and Exchange Commission, but these reports are of little value in managerial decision making.

Consider, for example, a machine that was purchased for a million dollars and that must, per tax regulations, be depreciated over a period of five years. To comply with these regulations, the cost accounting department records a depreciation of $200,000 each year and often assigns it as an overhead "cost" to production. The cost accounting system often wants the factory to make as many units as possible to reduce the overhead per unit, and the equipment-utilization metric is the result of this desire. The following real-life example shows how cost accounting metrics can easily become a suicide pact:

> To keep idle machine tools busy and consume fixed overhead, produced parts were being manufactured regardless of customer demand. Result: New bottlenecks were created and WIP built up in front of them as the unneeded parts moved downstream. The

inventory of unneeded parts grew, and the production of these parts competed for the same resources needed to produce customer-required parts. These unneeded parts consumed valuable machine time, scarce secondary processing resources, and assembly labor. This contributed to late deliveries.[16]

Toyota's Taiichi Ohno also recognized the waste of waiting and observed that this waste is often concealed—and aggravated—by the waste of overproduction:

> In any manufacturing situation, we frequently see people working ahead. Instead of waiting, the worker works on the next job, so the waiting is hidden. . . . In the Toyota production system, this phenomenon is called the waste of overproduction—our worst enemy—because it helps hide other wastes.[17]

Managers must realize that the investment in the tooling is really a *sunk cost* no matter how the cost accounting department must depreciate this investment. The manufacture of unnecessary inventory costs money. It does not, as many erroneously assume, recover it.

Sunk Costs

A *sunk cost* is money that is already gone for good or ill, and it merits a detailed explanation:

> A sunk cost exists because of actions taken in the past, not because of a decision made currently. Therefore, a sunk cost is *not* a differential cost. No decision made today can change what has already happened.[18]

Although the cost accounting system may have to account for this investment through depreciation, it has no relevance whatsoever to managerial decision making. Henry Ford explicitly noted that while the law required the valuation of plant and equipment according to accounting procedures (which Ford called "meaningless"), they were actually worth only what their owner(s) could do with them.[19] Taiichi Ohno said almost the same thing: A machine's value is defined solely by its current earning power.[20] Rigg adds:

> The mental static generated by sunk costs is hard to comprehend unless there is personal involvement. A large sunk cost can be detrimental to the current year's profit picture and may cause censure of the current managers, even if they had nothing to do with the original purchase and the depreciation schedule that caused it. There may also be a feeling that, "We've got so much invested in it that we can't afford to sell it for what it would bring."[21]

The last sentence immediately brings to mind the old adage about "sending good money after bad" and is a useful reminder that traditional cost accounting looks at where we were while managerial or engineering accounting looks at where we are going. No one would think of driving a car or piloting an aircraft by looking out the back window, but many business organizations have been driven or piloted into crashes through equivalent backward-looking financial metrics.

Overutilizing a million-dollar piece of equipment by distributing its depreciation costs over as many pieces as possible will simply waste even more money on the production of worthless inventory. If the tool is really of no use whatsoever, it should be sold to someone who can use it, even if it must be sold at a loss. If, for example, the highest bid for a million-dollar white elephant is $300,000, the owner's choice is not, "Should I keep this or suffer a $700,000 loss?" The real choice the owner must make is keeping something that is of no value or recovering $300,000 by selling it to the highest bidder.

Robert Sheckley's "The Laxian Key"[22] is an excellent science fiction story that illustrates the perils of overutilization. A man buys a machine called a Free Producer from Joe's Interstellar Junkyard and figures that he can make a fortune with it. He turns it on and it begins to produce a gray substance, apparently out of nothing. He soon discovers, however, that the machine is pulling energy from nearby power lines to create its product, and the power company soon demands that he pay for the electricity. The first lesson is that nothing is free, and it is the same in real-world factories. It costs money to utilize equipment because the materials are not free, either.

At this point, the story's protagonist naturally wants to turn the machine off but, upon reading the instructions for the first time, he discovers that he needs something called a Laxian Key. Since he doesn't have one, he puts the Free Producer in his space ship and departs for the home world of the aliens for whom the gray substance, Tangreese, is food. Upon his arrival, he is greeted by angry customs officials who demand that he remove the machine from the planet immediately. In desperation, he offers them the machine as a gift and urges them to feed their poor with it. The customs officials reply by pointing to endless hills of gray Tangreese. They reiterate their demand that he get the Free Producer off their planet while adding "But if you ever find a Laxian Key, come back and name your price!"

The problem in real-world factories is not the lack of an off switch but dysfunctional utilization metrics that prevent factory personnel from stopping a prodigal machine from wasting resources and producing inventory whose sole purpose is to absorb overhead or depreciation costs. It might, in fact, be useful to share "The Laxian Key" with plant personnel while referring to visible piles of inventory as Tangreese. This could result in a very different perspective of the inventory, which many people still think of as an asset because the accounting system treats it as one.

The Return on Investment (ROI) Metric

The sunk cost concept also applies to return on investment (ROI), another dangerous financial metric. In essence, ROI is the quotient of income, which is reduced by depreciation, divided by the amount invested in a piece of equipment:

$$ROI = \frac{Income - Depreciation}{Investment}$$

This is a strong argument—from the cost accounting perspective—to keep obsolete equipment running because the depreciation is zero and the book value, which goes in the denominator, also is zero. Unfortunately, the company owning the equipment is likely to end up wondering why its outstanding ROI did not prevent a more enlightened competitor (using state-of-the-art production equipment) from wiping it off the face of the earth. Any company wanting to avoid this fate would be wise to heed Henry

Ford's comment on the true value of plant and equipment: "Actually they are worth only what we can do with them."

VARIABLE COSTS IN THE THEORY OF CONSTRAINTS

Traditional cost accounting usually treats hourly labor as a variable cost. The standard labor cost for a unit of production is simply the hours of labor that are necessary to produce it, multiplied by the wage and benefit costs of each labor hour. This is eminently logical until one realizes that, in practice, the factory pays for nonovertime labor regardless of whether it uses it. Labor becomes a truly variable, or more precisely marginal, cost only when:

- The factory lays off workers when they are not needed and pays them only when they are.
- The factory is paying overtime.

Otherwise, only those things that actually go into the product are true variable costs. These always include materials and purchased parts. If substantial energy is required, as it is in the manufacture of many chemical products, energy also is a variable cost. For example, it takes twice as much electric power to process two tons of bauxite into aluminum as it does to process one ton. Factory lighting and heating are, on the other hand, fixed expenses (and sunk costs) that do not change with throughput.

A simple exercise in elementary calculus shows why fixed costs should be ignored in operational decision making. Profit equals revenue minus costs, and costs can be broken down into fixed and variable costs. Revenue is generally a function of the quantity produced and sold (n), and the same goes for genuinely variable production costs. This equation assumes, of course, that the factory does not produce anything that is not going to be sold, and the Theory of Constraints' performance measurements define throughput as finished goods with customers. For the purposes of the following discussion, n is defined as *throughput*. Then marginal and differential costs are defined as follows:

$$\text{Marginal cost of making another unit} = \left.\frac{d(Cost)}{dn}\right]_{n=N} = \text{material cost*}$$

$$\text{Marginal revenue from selling another unit} = \left.\frac{d(Revenue)}{dn}\right]_{n=N} = \text{sale price}$$

$$\text{Marginal profit from making and selling another unit} = \frac{d(\textit{Profit})}{dn}\bigg]_{n=N} = \frac{d(\textit{Revenue-Cost})}{dn}\bigg]_{n=N}$$

where N is the plant's current production level.

*and energy costs for energy-intensive products, along with overtime labor.

Revenue is a function $R(n)$ of throughput, and the same goes for variable costs $C(n)$. Fixed costs F include the producer's cost of capital, insurance, heating, lighting, and indirect staff. Profit is then calculated as $P = R(n) - C(n) - F$. Profit is maximized by setting the first derivative equal to zero as shown here:

$$\frac{Dp}{dn} = \frac{dR(n)}{dn} - \frac{dC(n)}{dn} - \frac{dF}{dn} = 0 \Rightarrow \frac{dR(n)}{dn} - \frac{dC(n)}{dn} = 0$$

The fact that the first derivative of a constant *(F)* is zero is obvious to any first-semester calculus student, but it is apparently not so obvious to executives who demand the production of excess inventory to "absorb overhead."

$R(n)$ and $C(n)$ are linear functions of n as long as the per-unit sale price and per-unit variable cost are constant. This means that profit cannot be maximized, and that the factory's objective should be to sell as many units as possible. In practice, of course, one of two things will happen as throughput increases:

- It will be necessary to hire more workers or pay overtime, in which case, $C(n)$ will become a step function that depends on n. If the per-unit material cost is M, the overtime labor cost is L, and K is the maximum possible throughput on regular time, $C(n) = nM + \max[0, (n - K)L]$.
- The factory will reach its capacity and no further throughput will be possible.

The following discussion shows how traditional cost accounting can result in highly dysfunctional decisions. If an item's bill of materials (BOM) calls for $8 in purchased materials, the differential or marginal cost of making another unit is $8.

Now suppose that, on paper, each unit also costs $4 in direct labor and $8 in allocated overhead, so the accounting book value of a finished piece is

$20. The plant has excess capacity, and a customer will buy units for $12. This is where the cost accounting system can become a suicide pact. The company may reject out of hand the idea of losing $8 per unit (perhaps with derisive jokes about selling at a loss and making it up on the volume). The truth is, however, that the company is paying for the labor and overhead whether it fills the order or not. The marginal or differential profit on each transaction is $12 (the sale price) minus $8 for the materials, or $4. Taking an order for 1,000 units under these conditions puts the company $4,000 ahead of where turning down the order would put it, and that is how managerial decisions should be made.

Now suppose that the plant has no excess labor, and filling the order will require overtime. Now the differential cost of making each unit is $8 for materials plus $6 ($4 at time-and-a-half) for labor, or $14. Twelve dollars is no longer an acceptable price under these conditions.

Other forms of overhead also are sunk costs, and Henry Ford explained this very explicitly:

> The foremen and superintendents would only be wasting time were they to keep a check on the costs in their departments. There are certain costs—such as the rate of wages, the overhead, the price of materials, and the like, which they could not in any way control, so they do not bother about them.[23]

The identification of the constraint is quite straightforward in a linear single-product process. It becomes more complicated, however, when the factory makes multiple products. The following section shows how to find the constraint by using linear programming.

FINDING THE CONSTRAINT

The Goal describes the hard way to find the constraint(s), as do Horace Lucien Arnold and Fay Leone Faroute:

> As soon as the rollways were placed, the truckers were called off, the floor was cleared, and all the straw boss had to do to locate the shirk or operation tools in fault, was to glance along the line and see where the rollway was filled up.[24]

In other words, "Look for a huge inventory backup somewhere in the factory, and the operation for which the inventory is waiting is the constraint."

The rollway was a means of transferring parts between workstations at Henry Ford's Highland Park factory. The work was supposed to be in continuous motion, and the appearance of inventory on the rollway between any two workstations showed exactly where the stoppage was. In Figure 1-3, there is either a stoppage at operation 3 or this operation itself is the constraint. In contrast, the single piece that is moving toward operation 2 will be used as soon as it arrives.

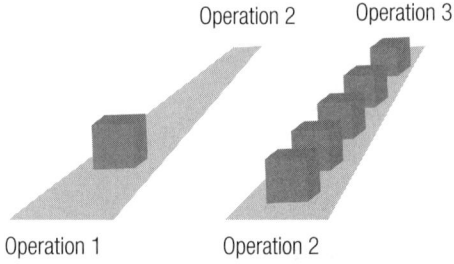

Figure 1-3. Discovery of the constraint (or stoppage).

The Goal has a similar description of how the constraint makes its presence obvious:

> "There's some of the proof," he says as he points to the stacks of work-in-process inventory nearby—weeks of backlog according to the report Ralph and Stacey put together and which we reviewed about an hour ago.[25]

Needless to say, searching for and finding a huge backlog of work-in-process is not the best way to identify the constraint. Identifying the constraint analytically, on the other hand, is fairly easy:

- In a single-product factory with unlimited demand for the product, the capacity-constraining resource is simply the operation with the smallest capacity. The only way to increase throughput is to elevate the constraint. This may be done by:

THE THEORY OF CONSTRAINTS

- Adding workers and/or more equipment
- Reducing non-value-adding setup time through single-minute exchange of die (SMED)
- Reducing non-value-adding downtime through preventive maintenance
• In a factory with limited demand for its products, the market is the constraint. To increase throughput, the factory must find more customers for the products as described in *The Goal*.
• In a factory with multiple products, the bottlenecks may depend on the product mixture. *Linear programming* is a standard operations research tool that can identify them. Linear programming can account for a combination of market and capacity constraints while identifying the product mixture that maximizes the differential profit.

Linear Programming

A common problem involves maximizing or minimizing an objective variable subject to constraints. These constraints may include the capacity of factory equipment as described in the preceding section, as well as market constraints and even contractual obligations to produce less profitable products. The following equation shows how a typical linear programming problem is formulated.

For n products and m constraints,

$$\text{Maximize } Z = \sum_{i=1}^{n} c_i x_i \text{ (Objective function)}$$

where the kth of m constraints is $\sum_{i=1}^{n} a_{ki} x_i \leq, =, \text{ or } \geq b_k$

where:
- x_i is the production quantity of the ith of n products
- c_i is the *marginal profit* from the ith of n products
- a_{mi} is the amount of the mth resource necessary to produce the ith product
- b_m is the availability of the mth resource (for example, equipment time or market demand)

In general, the constraints will be "less than or equal to." That is, the production plan cannot use more than any machine's available time, nor should it exceed the market demand for any product. The "greater than or equal" situation may apply, however, when the factory has a contract to produce a certain quantity of a product, or the factory wishes to do so to keep a customer happy. Application of linear programming will then yield the following information:

- The maximum possible marginal profit from the objective function
- The product mixture that is necessary to achieve this maximum
- The constraints (resources with zero excess or slack capacity) and the excess capacity that is available for the other resources
- The constraints' *shadow prices*, or differential benefits, for adding more of the resource

A capacity-constraining resource's shadow price is the marginal benefit of elevating that constraint. Nonconstraints do not have shadow prices because there is no benefit from adding more of a surplus resource. It is futile, for example, to ask the sales department to get more orders for a factory that is already working at capacity because the factory cannot fill the orders it already has.

Linear programming also allows the exploration of what-if scenarios, such as addition of capacity or changes in the product mixture. Elevating one constraint usually creates another, and linear programming is capable of finding this new constraint analytically.

Linear programming problems are solved by means of the *simplex method*, which the following section discusses.

Simplex Method

The linear programming problem must be rewritten as shown in the following equation, which is used for problems in which all resource constraints are "less than or equal." (The setup for "equal" and "greater than or equal" is more complicated.)

$$\text{Maximize profit } Z = \sum_{i=1}^{n} c_i x_i$$
$$\text{subject to } \sum_{i=1}^{n} a_{ij} x_i \leq b_j \text{ for all } m \text{ resources}$$

Rewrite this as:

$$Z - \sum_{i=1}^{n} c_i x_i = 0$$
$$\sum_{i=1}^{n} a_{ij} x_i + x_k = b_j \text{ (one equation per resource)}$$

x_k (where $k > n$) is the slack variable for the kth constraint or resource. It is the surplus or excess resource that remains for an intermediate or final solution in a linear programming problem. If the resource is machine capacity, the corresponding slack variable is the excess capacity. This will be zero by definition at the capacity-constraining resources (CCRs). Consider the example shown in Table 1-3.

Table 1-3. Product marginal profits and resource requirements.

Product	A	B	C	
Marginal profit per machine	$6	$6	$4	Available time in minutes
Drilling	$a_{11} = 4$	$a_{12} = 8$	$a_{13} = 6$	480
Grinding	$a_{21} = 6$	$a_{22} = 4$	$a_{23} = 4$	480
Heat treatment	$a_{31} = 8$	$a_{32} = 6$	$a_{33} = 10$	480

As the table shows, product A earns a marginal profit of $6, and it requires 4 minutes on the drill press, 6 minutes of grinding, and 8 minutes of heat treatment. The objective is to find the product mixture that maximizes the marginal profit, and also to identify the constraint(s). Set up the problem statement as follows:

$$Z = 6x_1 + 6x_2 + 4x_3 \text{ subject to}$$
$$4x_1 + 8x_2 + 6x_3 < 480$$
$$6x_1 + 4x_2 + 4x_3 < 480$$
$$8x_1 + 6x_2 + 10x_3 < 480$$

Rewrite this as:

$$Z - 6x_1 + 6x_2 + 4x_3 = 0$$
$$4x_1 + 8x_2 + 6x_3 + x_4 = 480$$
$$6x_1 + 4x_2 + 4x_3 + x_5 = 480$$
$$8x_1 + 6x_2 + 10x_3 + x_6 = 480$$

Table 1-4 is the initial *simplex tableau* for this problem. (B stands for basic variable and RHS for right hand side.)

A *basic variable* is one whose value is *not* zero, while a nonbasic variable is zero by definition. The initial simplex tableau begins with all the production levels (x_1, x_2, x_3) at zero; these are the nonbasic variables. The resources ($x_4, x_5,$ and x_6) are all 480 in this case, so they are the starting basic variables.

Table 1-4. Initial simplex tableau.

B	Z	x_1	x_2	x_3	x_4	x_5	x_6	RHS
Z	1	−6	−6	−4	0	0	0	0
x_4	0	4	8	6	1	0	0	480
x_5	0	6	4	4	0	1	0	480
x_6	0	8	6	10	0	0	1	480

This is obviously the worst-case situation; there is no production (and, hence, no marginal profit), and none of the resources have been assigned. The solution algorithm will eventually maximize the marginal profit, using all of at least one resource.

The next step is, therefore, to find the *entering basic variable*, which will replace one of the existing ones. The entering basic variable has the greatest positive effect ($-dZ/dx$) on the objective function; that is, increasing it maximizes the rate of increase in marginal profit. In this case, $-dZ/dx_1 = -dZ/dx_2 = 6$, so either of these variables may be chosen. This can, in rare cases, lead to a situation in which the solution procedure goes into an endless loop.[26] Rules are available for breaking ties, but it is usually acceptable to choose the entering basic variable arbitrarily. In this case, use x_1 so the column for x_1 becomes the *pivot column*.

The next step is to find the *leaving basic variable* and, hence, the *pivot row*. This is the one that has the smallest positive quotient after dividing the right-hand side (the remaining resource) by the pivot column element. In this case:

$$x_4: 480 \div 4 = 120$$
$$x_5: 480 \div 6 = 80$$
$$x_6: 480 \div 8 = 60$$

So x_6 is the *leaving basic variable*. Its row is the *pivot row*, and 8 is the *pivot element*. In simple English, what is happening is that the algorithm identifies the resource that will be exhausted most rapidly. Maximizing production for product A (represented in the table by x_1) will exhaust the capacity at heat treatment (x_6) before it exhausts the capacity of anything else. The next step is to divide the pivot row by the pivot element, as shown in Table 1-4a. Note that x_1 replaces x_6 as the basic variable.

Table 1-4a. Divide the pivot row by the pivot element.

B	Z	x_1	x_2	x_3	x_4	x_5	x_6	RHS
Z	1	−6	−6	−4	0	0	0	0
X_4	0	4	8	6	1	0	0	480
X_5	0	6	4	4	0	1	0	480
X_1	0	1	3/4	5/4	0	0	1/8	60

Then use standard matrix row operations to drive all the other numbers in the pivot column to zero, as shown in Table 1-4b. In this example, we multiply the pivot row by −6 and add it to the row directly above it (the row of slack variable x_5). Then for the indicated columns,

$$Z : 0 - 0 = 0$$
$$x_1 : 6 - 6 = 0$$
$$x_2 : 4 - \frac{9}{2} = -\frac{1}{2}$$
$$x_3 : 4 - \frac{15}{2} = -\frac{1}{2}$$
$$x_4 : 0 - 0 = 0$$
$$x_5 : 0 - 0 = 0$$
$$x_2 : 4 - \frac{3}{4} = -\frac{3}{4}$$

Table 1-4b illustrates that producing 60 units of product A ($x_1 = 60$) will yield $360 in marginal profit ($Z = 360$) while using all the heat treatment capacity. The drill press (x_4) has 240 minutes of excess capacity, and the grind-

Table 1-4b. Use matrix row operations to zero the other pivot column elements.

B	Z	x_1	x_2	x_3	x_4	x_5	x_6	RHS
Z	1	0	−3/2	7/2	0	0	3/4	360
X_4	0	0	5	1	1	0	−1/2	240
X_5	0	0	−1/2	−7/2	0	1	−3/4	120
X_1	0	1	3/4	5/4	0	0	1/8	60

ing station (x_5) has 120 minutes. This is not, however, an optimum solution as shown by the fact that there are still negative numbers in the Z row. As long as any coefficients for Z are negative, there is a positive dZ/dx available, which means that an increase in x will increase the marginal profit. Iteration must continue until all Z coefficients are positive or zero. Because the only negative coefficient is for x_2, x_2 will become the entering basic variable. What this means in practical terms is that it might be more profitable to make some product B at the expense of A. Now find the pivot row by dividing the right-hand side (surplus resource) by the coefficient in the pivot column.

$$x_4: 240 \div 5 = 48$$
$x_5: 120 \div -0.5$ does not qualify; consider only positive numbers
$$x_1: 60 \div 3/4 = 80$$

So x_4 will be the leaving basic variable. In other words, the new solution will exhaust the capacity of the drill press. Table 1-4c shows the tableau after dividing the pivot row by five to make the pivot element equal one. Fractions have also been converted to decimals.

Table 1-4c. Second iteration.

B	Z	X_1	X_2	X_3	X_4	X_5	X_6	RHS
Z	1	0	−1.5	3.5	0	0	0.75	360
X_2	0	0	1	0.2	0.2	0	−0.1	48
X_5	0	0	−0.5	−3.5	0	1	−0.75	120
X_1	0	1	0.75	1.25	0	0	0.125	60

In Table 1-4d, the other elements of the pivot column are driven to zero as before.

Table 1-4d. Second iteration, continued.

B	Z	X_1	X_2	X_3	X_4	X_5	X_6	RHS
Z	1	0	0	3.8	0.3	0	0.60	432
X_2	0	0	1	0.2	0.2	0	−0.1	48
X_5	0	0	0	−3.4	0.1	1	−0.80	144
X_1	0	1	0	1.10	−0.15	0	0.20	24

This is the optimum solution because the Z row contains no negative numbers. Also pay attention to the Z row's coefficients for the three slack variables: 0.3, 0, and 0.6 respectively for x_4 (drill press), x_5 (grinding station), and x_6 (heat treatment) respectively. Here is what it means in practical terms:

- The maximum marginal profit ($432) results from making 24 units of product A (x_1) and 48 of B (x_2).
- The grinding station (x_5) has 144 minutes of surplus capacity.
- The drill press (x_4) and heat treatment (x_6) have no surplus capacity so they are the capacity-constraining resources. Note that x_4 and x_6 are nonbasic variables because they do not appear in column B at the end of the iteration process. The same goes for x_3, which means that the optimum solution involves no product C.
- 0.3, 0, and 0.6 are the respective shadow prices for x_4 (drill press), x_5 (grinding station), and x_6 (heat treatment) respectively. Shadow prices are the marginal or incremental benefit from increasing a resource.

In this case, adding a minute of capacity to the drill press yields 30 cents of marginal profit, while another minute of heat treatment capacity adds 60 cents. Because the grinding station already has excess capacity, there is no benefit to be derived from adding more.

Shadow prices must be used carefully. In practice, adding a minute of drilling capacity will do nothing because, to begin with, no additional product can be made with less than four minutes of drilling. Secondly, there is no excess capacity at heat treatment. Suppose, for example, that 8 minutes of capacity are arbitrarily added to the drill press. The marginal profit from the optimum solution is $434.40, which is, indeed, $2.40 (8 × 0.30) more than $432. The problem is that the solution involves making 22.8 units of A and 49.6 units of B. It is, of course, impossible to make fractional products, but it should be possible to make 23 A units and 49 B units, for a total marginal profit of $432. As a practical matter, then, elevating the drilling constraint by itself does nothing.

Suppose, however, that 6 minutes of heat treatment capacity also are added. Then the optimum solution is to make 24 of A and 49 of B for $438—an increase of $6 in marginal profit. This does, in fact, match what is expected from the respective shadow prices: 8 × 0.30 (drilling) + 6 × 0.60 (heat treatment) = $6. The lesson is that shadow prices can theoretically indicate the best constraint to elevate, but that the resulting what-if scenario must be evaluated to make sure the solution is practical. Figure 1-4 shows a solution using the AB:POM (production and operations management) software from Jay Heizer and Barry Render.[27]

	x1	x2	x3		RHS		
maximize	6	6	4				
const 1	4	8	6	≤	480.00		
const 2	6	4	4	≤	480.00		
const 3	8	6	10	≤	480.00		

	x1	x2	x3	slk 1	slk 2	slk 3	RHS
maximize	−6.00	−6.00	−4.00	0.00	0.00	0.00	0.00
slk 1	4.00	8.00	6.00	1.00	0.00	0.00	480.00
slk 2	6.00	4.00	4.00	0.00	1.00	0.00	480.00
slk 3	8.00	6.00	10.00	0.00	0.00	1.00	480.00
Iteration 1	x1	x2	x3	slk 1	slk 2	slk 3	RHS
maximize	0.00	−1.50	3.50	0.00	0.00	0.75	360.00
slk 1	0.00	5.00	1.00	1.00	0.00	−0.50	240.00
slk 2	0.00	−0.50	4.00	−350	0.00	1.00	120.00
x1	1.00	0.75	1.25	0.00	0.00	0.125	60.00
Iteration 2	x1	x2	x3	slk 1	slk 2	slk 3	RHS
maximize	0.00	0.00	3.80	0.30	0.00	0.60	432.00
x2	0.00	1.00	0.20	0.20	1.00	−0.10	48.00
slk 2	0.00	0.00	−3.40	0.10	1.00	−0.80	144.00
x1	1.00	0.00	1.10	−0.15	0.00	0.20	24.00

Final	x1	x2	x3		RHS	Shadow
maximize	6	6	4			
const 1	4	8	6	≤	480.00	0.30
const 2	6	4	4	<	480.00	0.00
const 3	8	6	10	<	480.00	0.60
Values →	**24.00**	**48.00**	**0.00**		**$432.00**	

Figure 1-4. AB:POM solution.

- "slk" means "slack variable."
- slk 1, 2, and 3 are for drilling, grinding, and heat treatment respectively.
- The shadow prices have been circled in the final tableau.

The program's final report also identifies the shadow prices, along with the optimum solution and maximized objective function.

Because Microsoft Excel is widely available, the appendix shows how to set up and solve linear programming problems with Excel's Solver tool.

The Market Constraint

How does linear programming formulate a problem in which the marketplace is the constraint? Suppose that, in the preceding example, the demand for product C is unlimited, but customers will buy only 16 units of A and 32 units of B. The problem is formulated as follows:

$$\text{Maximize } Z = 6x_1 + 6x_2 + 4x_3 \text{ subject to}$$
$$4x_1 + 8x_2 + 6x_3 \leq 480$$
$$6x_1 + 4x_2 + 4x_3 \leq 480$$
$$8x_1 + 6x_2 + 10x_3 \leq 480$$
$$x_1 \leq 16$$
$$x_2 \leq 32$$

Figure 1-5 shows the AB:POM solution. The optimum solution produces 16 A, 32 B, and 16 C for a marginal profit of $352. The shadow prices show that heat treatment and the market demand for both A and B are the constraints, and that the greatest benefit would come from getting more orders for B ($3.60).

Problem statement

	A	B	C		RHS
maximize	6	6	4		
drilling	4	8	6	≤	480.00
grinding	6	4	4	≤	480.00
heat treat	8	6	10	≤	480.00
market	0	1	0	≤	16.00
market	0	1	0	≤	32.00

Solution

	A	B	C		RHS	Shadow
maximize	6	6	4			
drilling	4	8	6	≤	480.00	0.00
grinding	6	4	4	≤	480.00	0.00
heat treat	8	6	10	≤	480.00	0.40
market	0	1	0	≤	16.00	2.80
market	0	1	0	≤	32.00	3.60
Values →	16.00	32.00	16.00		$352.00	

Figure 1-5. Linear programming with market constraints.

Linear programming allows easy exploration of what-if alternatives such as cutting prices to get more orders. *The Goal* describes a scenario in which an offshore customer offers to buy a product for whose production there is surplus capacity, but the customer also wants a deep discount. Consider the preceding example, but assume that another customer will take up to 20 units of B for a one-dollar discount, thus reducing the marginal profit to $5. Figure 1-6 treats the discounted B as a separate product (B1) with identical resource requirements for its manufacture and a market constraint of 20 units. The discounted price deal will increase the marginal profit even if the fractional products must be rounded down to make the solution practical.

Problem statement

	A	B	B1	C		RHS
maximize	6	6	5	4		
drilling	4	8	8	6	≤	480.00
grinding	6	4	4	4	≤	480.00
heat treat	8	6	6	10	≤	480.00
market A	0	1	0	0	≤	16.00
market B	0	1	0	0	≤	32.00
market B1	0	0	1	0	≤	20.00

Solution

	A	B	B1	C		RHS	Shadow
maximize	6	6	5	4			
drilling	4	8	8	6	≤	480.00	.590909
grinding	6	4	4	4	≤	480.00	0.00
heat treat	8	6	6	10	≤	480.00	0.04545
market A	0	1	0	0	≤	16.00	3.27273
market B	0	1	0	0	≤	32.00	1.00
market B1	0	0	1	0	≤	20.00	0.00
Values →	16.00	32.00	14.545	7.2727		$389.8182	

Figure 1-6. What-if analysis of a price concession.

Contractual Requirements and Overtime

Linear programming can also model situations in which a factory *must* deliver a certain quantity of a less-profitable product to fill a contract or satisfy a customer. Suppose, for example, the company must supply 40 units of product 3. In this case, the constraint $x_3 \geq 40$ is added to the existing list of constraints.

The simplex method handles equality and greater-than-or-equal constraints by assigning an infinite penalty value (sometimes called the *Big M*) to solutions that fail to meet these requirements.

Hopp and Spearman show how to work overtime into the solution.[28] Recall that regular labor is usually a fixed cost even if workers are paid by the hour, whereas overtime is a marginal cost that offsets some of the marginal profit from making more items.

SUMMARY: THE THEORY OF CONSTRAINTS

This chapter has introduced the Theory of Constraints, which posits that no manufacturing process can work more quickly than its constraint or capacity-constraining resource works. Time lost at the constraint is lost forever, a fact that the *opportunity cost* concept reflects.

The Theory of Constraints recognizes only three performance measurements: throughput (finished goods with customers for them); inventory; and operating expense. This book recommends that cycle time be considered as an additional measurement. Operational decision making should be based on marginal or differential costs, revenues, and profits as opposed to accounting costs. Chapter 2 deals with production control.

Key terms introduced in this chapter:

capacity-constraining resource (CCR): The resource whose capacity limits the rate at which its system can function. Also known as the *constraint*.

constraint: Same as the capacity-constraining resource (CCR).

differential (cost, profit, or revenue): The incremental change in cash from producing an additional unit. *Differential cost* is usually the materials cost alone, unless overtime is being paid for labor. If material transformation involves substantial energy costs, as it does in the chemical process industries, energy is also a differential cost. *Differential revenue* is the incremental cash for making and selling an additional unit. It is usually the selling price. *Differential profit* is the incremental cash change for making and selling an additional unit, and it equals the differential revenue minus the differential cost. In the simplest case, this is the selling price minus the materials cost.

inventory: A key metric in the Theory of Constraints. Inventory consists of the money that is invested in things the company hopes to sell.

Although traditional cost accounting treats it as an asset, it really represents tied-up capital. Less is better.

marginal (cost, profit, or revenue): Same as the differential cost, profit, or revenue.

operating expense: A key metric in the Theory of Constraints. Operating expense includes expenditures whose purpose is to create throughput. Less is better.

opportunity cost: The cost of foregoing an opportunity, for example, to sell additional units. It refers not to money that is lost or spent (in which case it would show up on the books) but rather to money that the organization could have made by exploiting an opportunity. Opportunity costs are, therefore, often invisible to cost accounting systems.

sunk cost: Money that has already been spent, for good or ill. Sunk costs should be irrelevant to current decision making.

throughput: A key metric in the Theory of Constraints. Throughput consists of finished goods with customers (as opposed to finished goods that sit in warehouses in the hope of finding customers). More is better.

Endnotes

1. Aleksandr V. Suvorov (from Menning, "Train Hard, Fight Easy"), p. 00.
2. Franklin, *The Way to Wealth*, p. 12.
3. Franklin, *Poor Richards Almanac*, p. 10.
4. Ford, *Today and Tomorrow*, p. 114.
5. Seo, *Managerial Economics*, pp. 362–363.
6. Rigg, *Engineering Economics*, p. 500.
7. Ohno, *Toyota Production System*, p. ix.
8. Levinson and Rerick, *Lean Enterprise*, p. 112.
9. Harrington, "Looks Good on Paper."
10. Robinson, *Modern Approaches to Manufacturing Improvement*, p. 49.
11. Standard and Davis, *Running Today's Factory*, pp. 226–227.
12. Franklin, *The Way to Wealth*, pp. 20–21.
13. Ford, *My Life and Works*, pp. 144–145.
14. Ford, *Today and Tomorrow*, pp. 118–119.
15. Hopp and Spearman, *Factory Physics*, pp. 342–343.
16. Voiland, "A Nice Problem to Have," pp. 29–31.

17. Ohno, *Toyota Production System*, p. 59.
18. Anthony and Reece, *Accounting*, p. 712.
19. Ford, *Moving Forward*, p. 25.
20. Ohno, *Toyota Production System*, p. 64.
21. Rigg, *Engineering Economics*, p. 62.
22. Sheckley, *The Laxian Key*, p. 0.
23. Ford, *My Life and Work*, p. 98.
24. Arnold and Faurote, "Ford Methods and the Ford Shops," pp. 279–280.
25. Goldratt and Cox, *The Goal*, p. 144.
26. Hillier and Lieberman, *Introduction to Operations Research*, p. 0
27. Heizer and Render, *Production and Operations Management*, p. 0.
28. Hopp and Spearman, *Factory Physics*, Chapter 16.

TWO

Production Control: Pull versus Pig-Swallowing

The Theory of Constraints advocates a pull production control system known as *drum-buffer-rope* (DBR). Other pull systems like *kanban* (literally cards that serve as orders for more parts) can be used as well, and it is important to understand their inherent advantages over push systems.

Push-type systems force additional work into the production stream without regard for available capacity, as illustrated in Figure 2-1. A constrictor snake, like a python or boa, will swallow a pig and then sleep for a couple of months while it digests the intake. This results in the carriage of considerable inventory, although the snake's digestive system has evolved to handle it. This is not, however, true of most other animals (including humans) or manufacturing processes.

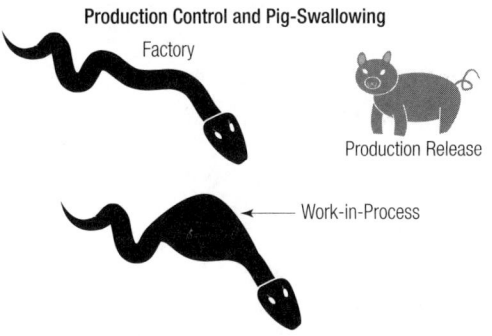

Figure 2-1. Push production control system.

Pig-swallowing is best left to systems that are designed to handle it (like pythons), but many factories build additional warehouse space to accommodate even more inventory. As described by Charles Standard and Dale

Davis, "Surprisingly, many factories prefer to 'stretch the python' so it can swallow an even larger hog!"[1]

As shown in Figure 2-2, the python cannot swallow anything else while the pig is occupying its stomach. Pig-swallowing similarly reduces the factory's responsiveness to new orders or else requires disruptive expediting—a form of industrial indigestion—to handle them.

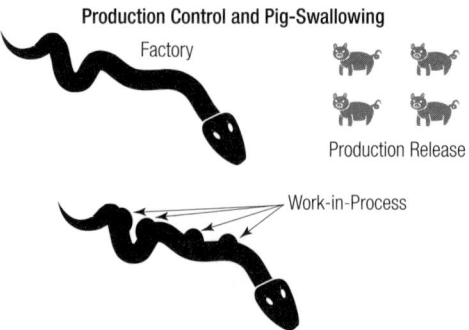

Figure 2-2. Push production works against responsiveness.

Figure 2-2 also introduces the concept of single-unit flow. The system might function more smoothly if it swallows four piglets—that is, small batches, instead of a large hog. This chapter will later show that the ideal production system approaches single-unit flow, and that batches are to be avoided if at all possible. The following section introduces the concept of just-in-time (JIT) manufacturing, a pull-type production system.

JUST-IN-TIME PRODUCTION CONTROL

JIT originated at the Ford Motor Company during the 1910s or even earlier, as described by Henry Ford:

> We have found in buying materials that it is not worthwhile to buy for other than immediate needs. We buy only enough to fit into the plan of production, taking into consideration the state of transportation at the time. If transportation were perfect and an even flow of materials could be assured, it would not be necessary to carry any stock whatsoever. The carloads of raw materials would arrive on schedule and in the planned order and amounts, and go from the railway cars into production. That would save a great deal of money,

PRODUCTION CONTROL: PULL VERSUS PIG-SWALLOWING

for it would give a very rapid turnover and thus decrease the amount of money tied up in materials. With bad transportation, one has to carry larger stocks.[2]

This statement illustrates the following key aspects of an ideal JIT production system:

- Materials are purchased only as they are needed.
- Delivery of materials is synchronized with production and, ideally, goes from the receiving dock directly to the point of use. (The receiving dock will, in fact, be near the point of use if possible to minimize in-plant transportation.)
- Inventory ties up working capital and should therefore be minimized.
- The ability to receive delivery in JIT-sized quantities depends on the reliability of the freight management system (FMS). If the freight runs like clockwork—and Ford designed his FMS to do exactly that—there is no need to carry buffer inventory for protection against factory-stopping stockouts. This relates directly to the issue of variation in material transfer times.

Ford's production system did not use kanban or a similar production control system; it was, as shown in the following section, designed to run like a clock.

Synchronized Production and Takt Time

The speed of the work was carefully timed so that the assembly line did not run too fast or too slow. Where the workers put together the chassis, the line moved 6 feet per minute. Where the workers bolted the front axle to the chassis, the line moved faster, 15 feet per minute It was like setting the mechanism of a clock. Henry had created a giant moving timepiece.[3]

Ford's assembly line was not a single assembly line but rather a set of lines that converged the way streams join a river. Another way to imagine the Ford production system is to think of a hierarchical bill of materials (BOM) in which each item's production *rate* matches its *proportion* in the downstream products. One line made chassis, another engines, and all worked at an apparently synchronized rate so there was rarely an excess or shortage of anything. The system really did have to work like a clock, and this suggests the concept of *takt time*—working to the rate of customer demand.

$$\text{Takt time} = \frac{\text{Available time}}{\text{Demand (units)}} = \frac{\text{time}}{\text{unit}}$$

Ford suggests that his factories did use the takt time concept:

> The speed of the moving work had to be carefully tried out; in the flywheel magneto, we first had a speed of sixty inches per minute. That was too fast. Then we tried eighteen inches per minute. That was too slow. Finally, we settled on forty-four inches per minute. The idea is that a man must not be hurried in his work—he must have every second necessary but not a single unnecessary second.[4]

Although highly complicated bills of material add difficulty to production planning, they do not make the problem insurmountable. Robert H. Hayes, Steven C. Wheelwright, and Kim B. Clark describe how Convair handled the production of the B-24 Liberator bomber: "Production control developed systems for timing the arrival of over 500,000 parts, subsystems, and systems for final assembly—without computers."[5] It is more than likely, in fact, that Ford's production systems played a major role in achieving this. Ford's production chief, Charles Sorensen, designed the Willow Run bomber plant, which was capable of producing one Liberator every hour:

> To compare a Ford V-8 with a four-engine Liberator bomber was like matching a garage with a skyscraper, but despite their great differences I knew the same fundamentals applied to high-volume production of both, the same as they would to an electric egg beater or to a wristwatch. First, break the plane's design into essential units and make a separate production layout for each unit. Next, build as many units as are required, then deliver each unit in its proper sequence to the assembly line to make one whole unit—a finished plane.[6]

PRODUCTION CONTROL: PULL VERSUS PIG-SWALLOWING

The concept of rhythmic timing is actually thousands of years old. One reason the Greek phalanx was so lethal was that the hoplites (heavily-armored spearmen) marched in step, a process undoubtedly facilitated by musical instruments like flutes and drums. The phalanx could not operate properly if the faster soldiers got ahead of the slower ones; it would lose both the protection of overlapping shields and the advantage of having everyone hit the enemy simultaneously. (This chapter later shows how the drum-buffer-rope production control system uses the same concept, but it also applies to the choreographing of paced assembly lines that must work in synchronization like the gears and wheels of a mechanical clock.)

After the introduction of gunpowder, volley tactics required every soldier to load and fire at the same rate. Baron von Steuben's *Regulations for the Order and Discipline of the Troops of the United States* called for the soldiers to mark one second of time between each carefully prescribed motion of the loading drill, presumably to keep everyone working at the same rate. Sailors, meanwhile, used songs to provide rhythm for group tasks, such as hauling on ropes. "On many occasions the Shantyman would play an instrument (usually a fiddle) and sit in the middle of the capstan while the sailors heaved with synchronized teamwork to the Shantyman's music and song."[7] As described by Richard Henry Dana, Jr., in *Two Years Before the Mast* (emphasis added),

> The sailors' songs for capstans and falls are of a peculiar kind, having a chorus at the end of each line. The burden is usually sung, by one alone, and, at the chorus, all hands join in—and the louder the noise, the better. With us, the chorus seemed almost to raise the decks of the ship, and might be heard at a great distance, ashore. *A song is as necessary to sailors as the drum and fife to a soldier. They can't pull in time, or pull with a will, without it.* Many a time, when a thing goes heavy, with one fellow yo-ho-ing, a lively song, like "Heave, to the girls!" "Nancy oh!" "Jack Crosstree," etc., has put life and strength into every arm. *We often found a great difference in the effect of the different songs in driving in the hides.* Two or three songs would be tried, one after the other, with no effect; not an inch could be got upon the tackles—when a new song, struck up, seemed to hit the

humor of the moment, and drove the tackles "two blocks" at once. "Heave round hearty!" "Captain gone ashore!" and the like, might do for common pulls, but in an emergency, when we wanted a heavy, "raise-the-dead" pull, which should start the beams of the ship, there was nothing like "Time for us to go!" "Round the corner," or "Hurrah! hurrah! my hearty bullies!"

Herman Melville's *Redburn: His First Voyage* adds,

But I soon got used to this singing; for the sailors never touched a rope without it. Sometimes, when no one happened to strike up, and the pulling, whatever it might be, did not seem to be getting forward very well, the mate would always say, "*Come, men, can't any of you sing? Sing now, and raise the dead.*" And then some one of them would begin, and if every man's arms were as much relieved as mine by the song, and he could pull as much better as I did, with such a cheering accompaniment, I am sure the song was well worth the breath expended on it. It is a great thing in a sailor to know how to sing well, for he gets a great name by it from the officers, and a good deal of popularity among his shipmates. Some sea-captains, before shipping a man, always ask him whether he can sing out at a rope.

Music can also be used to coordinate some very complicated activities, such as ballets in which different groups of dancers perform different routines that must take place together. Marching bands that turn themselves into complex formations during football halftime performances are another example. This suggests that it is quite possible to choreograph a set of manufacturing processes, especially when automation eliminates all human variation from processing and transportation times.

The fact that Ford's production system could be choreographed so finely that there was rarely a shortage or excess of anything, even when the plant was running at almost 100 percent capacity, makes its technology worthy of further study. It was not, however, a true pull system because downstream operations did not control the pace of upstream operations.

PRODUCTION CONTROL: PULL VERSUS PIG-SWALLOWING

Kanban Systems

Kanban means "card," and a kanban system uses visual controls like cards or empty containers to signal upstream operations to produce more parts. When workstations are within sight of one another, delivery points or kanban squares can be used instead (Figure 2-3).

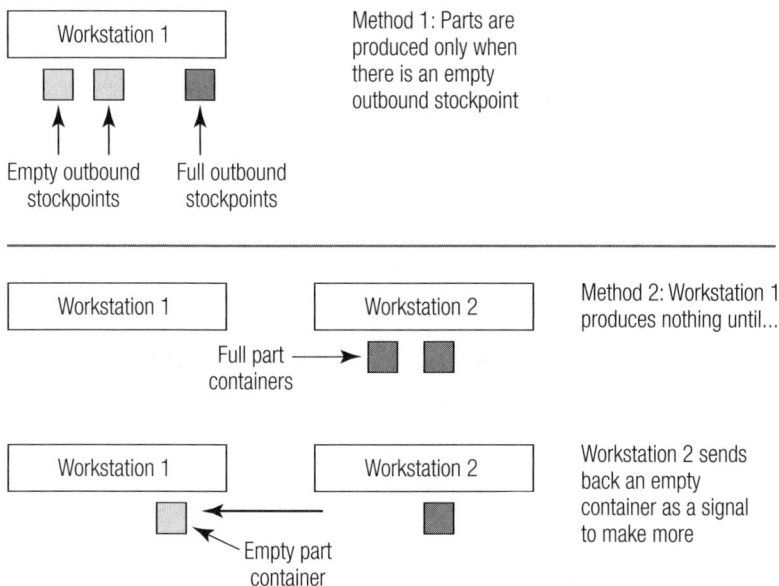

Figure 2-3. Kanban-type production control examples.

Systems of this nature were, in fact, in use in the United States at the beginning of the 20th century. The system described below actually combines aspects of kanban systems and visual controls like andon lights that indicate the status of production equipment.

A red signal card is put in the top compartment in the planning department route rack of each machine not in operation.... If there is no work for the machine to do, the condition of the rack shows it by the absence of next-work order forms in either the middle or top compartment. A red signal card in a machine compartment with work orders in the middle compartment, but none in the top compartment, indicates that the machine is without an operator and that no work is set in it. If there is a work order in the top compartment behind the red signal card, it indicates that the machine is without an operator and the job specified on the said work order is standing idle in the machine. If the red signal card in any of the above cases displays a large letter R it indicates that the machine is undergoing repairs and cannot yet be operated. The absence of a red signal card indicates that the machine is running.[8]

This visual control system was, therefore, capable of indicating the following conditions:

- The equipment is idle due to lack of work.
- The equipment is idle due to lack of operator.
- The equipment is down for repairs.
- The equipment is acting on a job.

If the machine is a capacity-constraining resource, neither of the first two conditions listed above can be allowed to happen.

Figure 2-4[9] shows a visual control that has the same effect as a kanban system. As there are only three metal pockets at each machine, and one of these is occupied by the instruction card for the job the machine is working on, it is clearly impossible to add more than two jobs to the queue. The effect is, in fact, identical to that of having two kanban squares at the machine; if both are occupied by waiting jobs, no more are sent.

Another factory devised a system that seems to improve on the kanban square (which must be empty before new work can be placed into it). If a worker predicted that a tool was about to become ready for the next job in a queue, he or she sent an audible signal to pull work from the appropriate source, ensuring the job would arrive just in time—perhaps seconds or minutes early to avoid stoppages.

> A large blacksmith shop has cleverly solved the problem of keeping the workmen supplied with jobs without loss of time to the pieceworker. Near each shear or cut-off in the shop is a push button connected with an indicator in the department office. A few minutes before the shear man is through with a job, he pushes a button, which indicates to the storekeeper that shear No. 12, for example,

PRODUCTION CONTROL: PULL VERSUS PIG-SWALLOWING

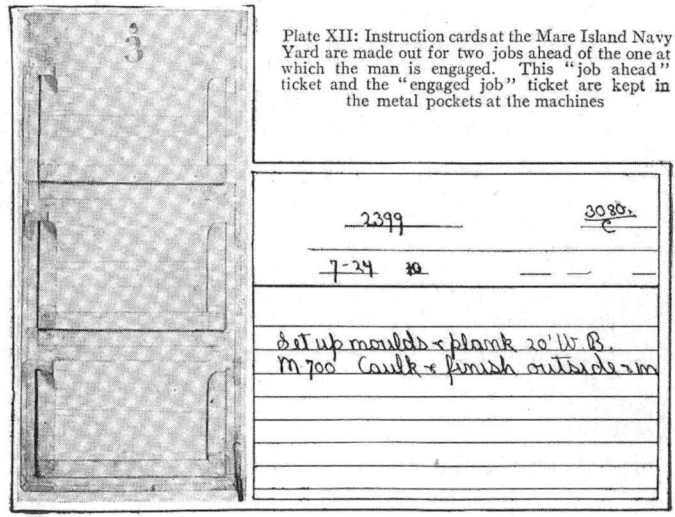

Figure 2-4. Kanban-like system at the Mare Island Navy Yard, early 20th century.

will soon be ready for stock for another job. The storekeeper then pushes a second button, which calls the man who has charge of the iron house. To him is given the original memo order, and he delivers the material to shear No. 12, in ample time to keep the workman busy, and prevent any loss of time on the shear.[10]

If this kind of pull production can be achieved with buzzers, it is easy to envision what might be achieved with modern computers and sophisticated kanban production systems. It is important to note, however, that all kanban production systems (simple or elaborate) have a common feature. Nothing can be produced without a kanban (or empty container, empty delivery point, or similar signal). A factory's inventory is, therefore, limited to the kanban population. If there is no empty container, empty kanban square, or similar request for parts, the parts are not produced.

The Ford production system often sped up the production line for the express purpose of exposing problems. If the workers couldn't keep up with the increased speed, an underlying cause was often revealed, and its correction allowed faster rates without further trouble. The Japanese will often remove kanban cards or containers (that is, protective inventory) or workers from the line for the same purpose. The idea is to see whether the factory can operate with even less inventory or fewer people and, if it can't, to identify potential improvement projects. Another way of looking at this is to imagine the factory as a stream in which water depth is protective inventory and rocks are variation sources. Reduction of the water depth (safety stock) makes the rocks visible, and they can be removed.

Hopp and Spearman provide a caveat for using kanban systems in a capacity-constrained factory.[11] It has already been pointed out that time lost at the constraint is lost forever, and the capacity-constraining resource must therefore be kept as busy as possible. Suppose, however, that the workstation that follows the constraint gets behind for some reason or other (even though it has nominally greater capacity) and stops sending kanban cards or empty containers to the constraint. The rule under kanban production control is to make parts only in response to a kanban, so the constraint becomes *blocked*; it is not allowed to produce anything. This suggests that drum-buffer-rope, the production control system of synchronous flow manufacturing, is preferable to kanban when a constraint is present. DBR ties the constraint or capacity-constraining resource to production starts.

DRUM-BUFFER-ROPE

Previous sections in this chapter described how actual drums synchronize complicated activities such as marching in step. The idea is that faster people (or activities) do not get out of step with slower ones. A drum-buffer-rope production control system has the following basic characteristics, as shown in Figure 2-5.

- The CCR beats the drum to set the pace for the entire factory.
- An information rope ties production starts to the CCR.
- An inventory buffer at (or heading toward) the constraint protects it from upstream stoppages.

The inventory buffer's purpose is, of course, to protect the CCR from unforeseen or uncontrollable variation in upstream production rates. More variation means that a larger buffer is necessary. The variation reduction chapter shows how to reduce the variation, and hence the required buffer, in a DBR system.

PRODUCTION CONTROL: PULL VERSUS PIG-SWALLOWING

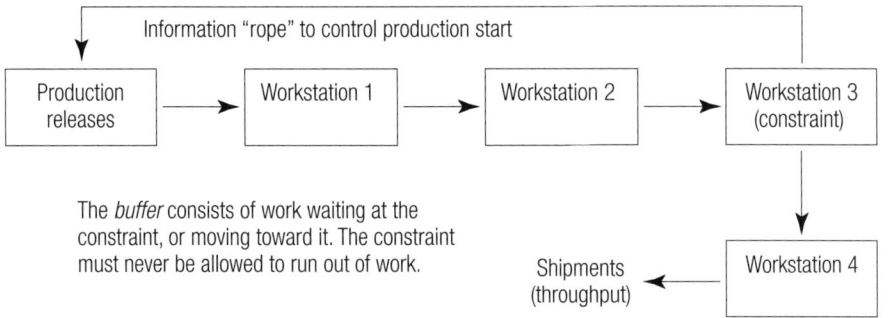

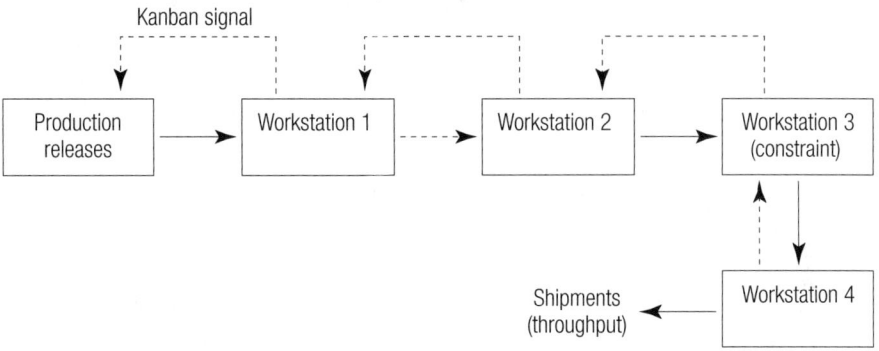

Figure 2-5. Drum-buffer-rope production control system.

Buffer Management

The inventory *buffer* is a necessary evil because any stoppage at the CCR is worse than carrying inventory. If the factory could rely on a steady flow of work to proceed toward the constraint without interruption, it would not have to maintain a buffer. This takes us back to the concept of variation that often creates a feast-or-famine situation at the CCR. The only way to reduce the buffer is to reduce the variation that makes it necessary. Wayne Smith, in fact, provides an equation for "product wheel" safety stock that underscores this point:[12]

$$\text{Safety stock} = Z\sqrt{\sigma_D^2 R^2 + \sigma_R^2 D^2}, \text{ where}$$

- R = replenishment time
- D = sales demand rate
- σ_R = standard deviation, replenishment time
- σ_D = standard deviation, sales rate
- Z = service rate coefficient or coefficient of risk (standard normal deviate for the probability of being on time; for example, 1.645 for 95%)

Smith additionally notes that the formula applies for single-function kanban or for multiple product kanbans. The key observation is that shorter replenishment times and less variation in replenishment times reduces the necessary safety stock.

Figure 2-6 shows a typical buffer management scheme reflecting Goldratt's approach, which divides the buffer into three zones: OK; Watch and Plan; and Act. The Act zone represents the work closest to the constraint, and any missing job will cause the constraint to run out of work unless jobs from the Watch and Plan and/or OK zones are expedited to make up the shortage.

Buffer management can, in fact, be compared to project management, in which any delay on the *critical path* (as identified by the Critical Path Method, CPM), will delay completion of the project. Goldratt, in fact, adapts DBR to project management in the form of Critical Chain.

In summary, DBR is a pull system like kanban. Its production control system differs from kanban because, instead of pulling work from one workstation to another, it pulls production starts into the system to replenish the buffer. To put this another way, kanban and related systems pull work from the operation directly upstream to fill an empty container, kanban square, or stock point. DBR pulls work into the line to fill empty spaces in the buffer. Although kanban systems require communication (often in the form of simple visual controls) between all adjoining operations, DBR has the advantage of requiring only one information "rope" to connect the CCR to production starts.

PRODUCTION CONTROL: PULL VERSUS PIG-SWALLOWING

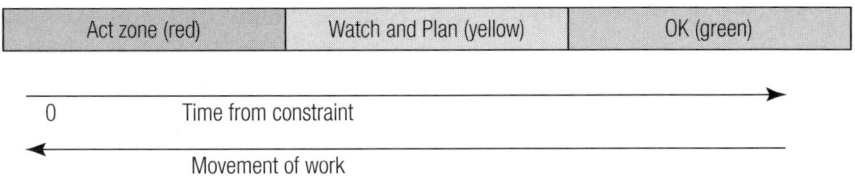

Example 1: A zone is full if it contains four units. This situation is acceptable because upstream workstations can easily make up the shortfall in the OK zone long before it reaches the constraint.

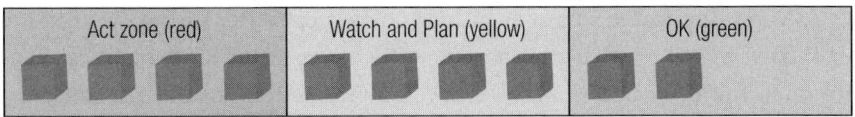

Example 2: This situation is very dangerous because a shortfall is about to reach and idle the CCR. The production control system must expedite work from an upstream zone to fill the "hole" before it reaches and stops the constraint.

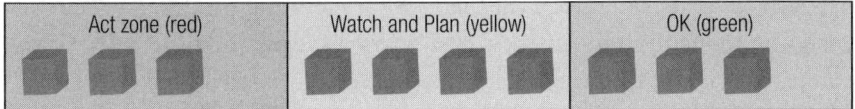

Figure 2-6. Buffer management.

Simplified Market Pull

When marketplace demand is the CCR, *simplified market pull* (SMP) can be used to control release of work to the shop floor. Dave Turbide explains that SMP "... implements easily and inexpensively, with minimal changes to normal business processes, and delivers outstanding results very quickly. Early adopters have cut work in process inventory in half, reduced lead times by 50 percent or more, and improved on-time completion to 98 percent plus—all within a few weeks of getting started with SMP."[13]

SMP's key features are (1) the order release logic and (2) the workflow system. Work must be released to the shop floor early enough to meet customer orders but not so early that it increases the WIP. This is simply a basic principle of DBR except that the market, and not a workstation, sets the pace for the line.

Meanwhile, the workflow system prioritizes orders. Make-to-order calculates an order's priority, depending on the remaining lead time and the time required to complete the order. According to Turbide, the logic is similar to the critical ratio prioritization rule. (Critical ratio = time remaining divided by hours of work remaining. As long as it is greater than one, the job is early. If it becomes less than one, the job is late.) For replenishment jobs, the priority depends on the ratio of the actual stock level to the customer's desired stock level.

SMP is applicable to any plant with excess capacity. Turbide points out that many plants that are running at only 75 percent to 80 percent capacity are often full of unnecessary WIP because production control releases jobs early to ensure that they are finished on time. (From the lean manufacturing perspective, "too soon" can often be as bad as "too late.") More WIP means longer cycle times, and this leads to a self-fulfilling prophecy: Jobs must be started earlier and earlier to make sure they are finished on time.

SMP actually provides a mechanism for elevating the (market) constraint by reducing lead times. If customers want just-in-time delivery, the company that can fill orders on time—but not too early—will have an advantage over its competitors. Shorter lead times, therefore, become a sales tool.

CONWIP

Hopp and Spearman[14] describe a CONWIP (Constant WIP) pull production control system, which is similar to drum-buffer-rope. As with DBR, there is only one information stream, as opposed to kanban links between all workstations in the sequence. CONWIP, however, ties production starts to withdrawals from the finished goods inventory instead of a constraint.

CONWIP may emulate kanban by using cards to trigger production, but the cards are not part-specific. They tell the factory *when* to release work into the line, but not *what* to release. The backlog of orders defines the sequence in which specific parts should be started.

CONWIP also seems to tie in with simplified market pull when the marketplace, and not a factory operation, is the constraint. If the constraint is in the factory, it must be kept busy because time lost there is lost forever. Therefore, the capacity-constraining resource should set the pace for production starts. If, however, the factory has excess capacity, it makes sense to let the market set the pace by tying production starts to depletion of the finished goods inventory or, even better, to specific customer orders.

No matter what kind of pull production system is used, flow variation creates the need for protective inventory buffers while limiting the system's productive capacity. This ties in with Hopp and Spearman's discussion of statistical throughput control.[15]

Statistical Throughput Control

Statistical process control (SPC) uses control charts to detect undesirable changes in a manufacturing process's mean or standard deviation. Statistical *throughput* control (STC) is similarly designed to detect impending stockouts to allow management to order overtime or take other corrective actions. It can probably be adapted for buffer management purposes. STC also helps raise awareness of variation in material transfer and processing times. Quality practitioners and manufacturing engineers are very familiar with SPC, and they understand that variation in product characteristics can cause nonconformances. The term "statistical throughput control" extends this familiar concept to the kind of variation that increases cycle times and inventory levels. Hopp and Spearman define the following parameters for use in STC:[16]

- R = time period in which a quota of Q items is to be completed
- n_t = cumulative production at time t (where we are)
- S_t = scheduled cumulative production at time t (where we should be)

If the production rate is to be steady, $S_t = Q\frac{t}{R}$. That is, if half the time is remaining, half of the quota should have been produced.

- z_α = the standard normal deviate that corresponds to a 100α percent chance of falling short of the quota
- μ = mean throughput over a time period of R
- σ = standard deviation for throughput over a time period of R

Once these values have been established, it is possible to create a statistical throughput control chart by means of the following equation:

$$x = -(\mu - Q)\frac{(R-t)}{R} - z_\alpha \sigma \sqrt{\frac{R-t}{R}}$$

The first term is the difference between the expected throughput and the quota, times the fraction of time remaining in period R. The second term is the $(1 - \alpha)$ quantile of the throughput that will actually occur. Figure 2-7 shows an STC chart for a production quota of 1,000 pieces in 10 hours, where the average throughput is also 1,000 pieces, and the standard deviation is 100 pieces.

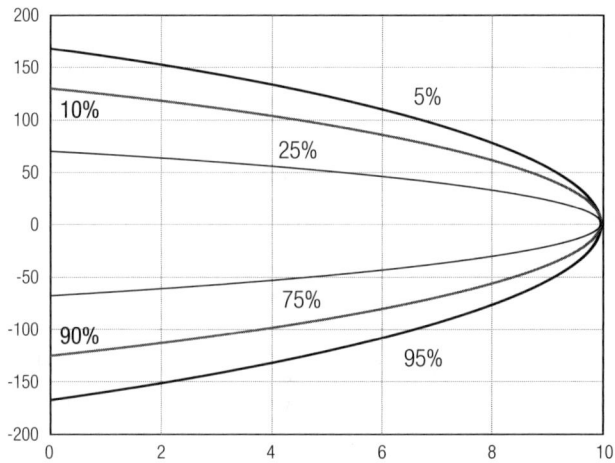

Figure 2-7. Statistical throughput chart.

The curved envelopes show the chance of falling short of quota given a current overage. For example, if the factory is 100 pieces ahead of quota after four hours, there is about a ten percent chance of falling short at the end of the allotted production period. Another way of saying this is that 500 pieces have already been produced, and the factory must make another 500 in the next six hours. The expected throughput for six hours is 600 pieces, and the standard deviation is 100 times the square root of 0.6 or 77.5. Calculate the standard normal deviate, as follows, to find the chance of making 500 or fewer pieces during the remaining time. The cumulative normal probability for −1.29 is 0.0985 or slightly less than ten percent.

$$z = \frac{\text{(number required)} - \text{(number expected)}}{\sigma} = \frac{500 - 600}{77.45} = -1.29$$

PRODUCTION CONTROL: PULL VERSUS PIG-SWALLOWING

Figure 2-8 simulates actual production with N normally distributed random numbers with mean μ/N and standard deviation $\sigma N^{-0.5}$. That is, if production period R is broken up into N segments, each of duration R/N, the mean production for those periods will be μ/N, and the standard deviation will be $\sigma N^{-0.5}$. It is quite clear that results that range from well above to well below quota can be obtained.

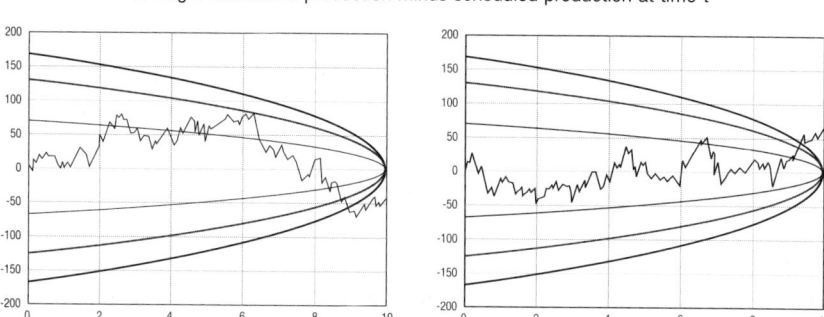

Figure 2-8. STC simulations.

In practice, of course, throughput should not fall short of quota. The purpose of an STC chart is to allow management to take corrective action, such as increasing the rates for preconstraint processes, to prevent shortfalls from ever taking place.

The following section introduces the concept of flow, and particularly continuous flow. The ideal is to keep the work in continuous motion. Chemical plants that deal with materials that flow and pour can, in fact, achieve such flow, but discrete-product factories can only seek to approximate it.

FLOW: KEEP THE WORK MOVING

To understand the concept of flow, it is useful to examine the way chemicals are manufactured in the chemical process industry (CPI). There is a huge difference between the test tubes and flasks that are found in chemical laboratories and the kind of equipment that is used for large-scale chemical manufacturing.

Chemical plants sometimes use batch reactors, which are essentially "flasks" that can be as large as several thousands of gallons. The batch reactor is the hardest reactor to control. As an example, it may be necessary to add or withdraw enormous quantities of heat during the initial stages of the reaction, when the reactants are most concentrated. Batch-to-batch variation is expected as a matter of course. Such reactors are often useful for small or specialty jobs, just as factories with departmental or job-shop layouts are good at handling relatively small jobs.

In all cases, it is important to keep a product moving, but the systems employed (*flow* lines or *paced* lines) for this purpose vary in function. At times, one or the other may be selected to accommodate management considerations. As the discussion below illustrates, it is important to distinguish between *flow* lines and *paced* lines and the relative advantages of each.

Flow Lines and Paced Lines

Hopp and Spearman underscore important differences between flow lines and paced lines.[17] In flow lines, each workstation operates at its own rate, and the overall rate is (or should be) set by the bottleneck or capacity-constraining resource. The drum-buffer-rope or kanban production control system can keep the faster workstations from getting ahead of the slower ones. It is normal and even desirable to have excess capacity at some workstations.

In a paced assembly line, the rate is set by the conveyor belt that carries the parts from one position to the next. In this case, the rate must be set so that each person or machine can keep up with the work. The ideal condition, as described by Henry Ford, is for each person to "have every second necessary but not a single unnecessary second."[18] This leads to the need to balance the paced assembly line. If the work is divisible into tiny increments, it can be assigned in roughly equal portions to each person or workstation. This is sometimes known as the line-of-balance (LOB) problem, for which Hopp and Spearman provide an algorithm.[19]

The Plug Flow Reactor: A Model for the Ideal Situation

Chemical engineers prefer the continuous stirred tank reactor (CSTR) and plug flow reactor (PFR) over the batch reactor. In a chemical plant, where capacity is limited by the unit with the lowest capacity (either the reactor or a downstream unit, like a separator or distillation column), the plug flow chemical reactor works in much the same way as a flow line, and the PFR (Figure 2-9) is a useful model for illustrating the kind of conditions that a discrete-product factory may wish to approximate. Richard J. Schonberger notes that the ideal state "is not assembly-line production but continuous production, which is found in what are often known as process industries."[20]

Differential fluid element

Figure 2-9. Plug flow reactor.

In a PFR model, there is essentially no variation in processing or material transfer times. Although the model cannot, of course, be used for discrete products, it is a useful paradigm for envisioning an ideal situation.

In a properly functioning PFR, there is no axial mixing between the infinite number of circular fluid elements (each of differential length) that are in continuous motion through the pipe, which may contain a catalyst. The fluid element is analogous to a miniature batch reactor that contains a single chemical *workpiece*. It is continuous motion through the pipe or tube, and a value-adding chemical reaction takes place during its entire progression through the tube. PFRs are relatively easy to control; they can, for example, be jacketed with steam or cooling jackets as needed.

At Henry Ford's River Rouge plant, workers stood on conveyors that moved with large items such as chassis. This reduced non-value-adding transportation time between one workstation and another and also made it unnecessary for a worker to move his body in time to a conveyor belt:

> Thus a workman is not obliged to divide his attention between his legs and his work. He rides with the job. When he completes it, he steps off the walk, takes a few paces to the rear, meets the next car requiring his attention, returns to the walk, and again goes to work![21]

The concept has been extended even further in modern automotive assembly plants with *skillets*, or moving platforms that hold the workers and the vehicles. As Burke Brown of DaimlerChrysler observed, "Ergonomically, it is much better for the worker because he can alter the height of the skillet, and is not forced to keep up with the vehicle as it travels down the assembly line."[22] Ford's moving "sidewalks" and the modern skillet are examples of how a discrete-product factory might try to approximate the ideal conditions that exist in a plug flow reactor:

- The work is in constant motion, with no stopping or sidelining as inventory.
- Value is added constantly, without pause even for transfer from one workstation to another. In the PFR, this concept is literal because the chemical reaction is in fact continuous. In a discrete-product situation, value will probably not be added continuously but the proportion of value-adding "bang!"—Masaaki Imai's description of the relatively short moment in which a tool makes contact with the part and transforms it—to non-value-adding waiting and transportation time should increase.
- The job is a single unit: a differential fluid element in a reactor, or a single discrete piece on an assembly line.

The Flying Boxes: Flow at the Start of the 20th Century

The chemical process industries are well aware of the virtues of getting materials to flow, and the pneumatic conveyor is a tool for making solids flow like liquids or gases. The typical application involves powders or pellets, such as plastic pellets as feedstock for extruders. One firm, however, applied the concept to paper boxes:

> A Detroit paper box concern uses an exhaust fan for carrying its product from one department to the next through sheet iron ducts. In the packing room are long rows of tables at which stand the packers. Before each employee, a spout leads down from the main duct to the packing table. There is a damper at each spout controlled by the operator in the storeroom [sic]. When the stock gets low on the packing table, the packer signals to the storeroom, the operator opens the damper and more boxes flow to the table. Enough can be fed in a few minutes to last an hour or more.

PRODUCTION CONTROL: PULL VERSUS PIG-SWALLOWING

[Because the boxes can be sent on demand, however, there does not seem to be a real reason to keep an hour's worth of inventory.] This method of transferring boxes saves the usual expensive trucking and does the work swiftly, nor are the goods injured by this method of handling.[23]

The box example suggests that pneumatic conveying might be suitable for making anything "fly" from one part of the factory to another, as long as it is relatively light and unlikely to be damaged in the process. Pneumatic tubes (such as those used by banks to move money and paperwork) are another example, but they involve containers that must at some point return to their points of origin.

The example, which comes from "Keep the Product Moving" (a chapter of a work published in 1911), also provides an interesting historical perspective on the concept of flow. As with many of these almost-centenarian examples, *the thought process is more important than the specific application.* The previous excerpt and the one below illustrate that American manufacturers had clearly recognized the concept almost 100 years ago. The same source states:

Keep material in process on trucks [hand trucks or carts] and off the floor. This will keep the shop clean and the parts moving. This axiom was proved again at the garment factory of the Hilker and Wiechers Company at Racine, Wisconsin. Garments when finished, before being sent to the storeroom, were piled on a bench in the sewing room. In taking them to the storeroom the boy brought an empty truck up the elevator, loaded the garments from the bench and unloaded them again on the stock room shelves.

Trucks suitable for carrying the garments were built. Empties stand always ready for the garments as they are finished and as fast as a truck is filled, all the boy has to do is to keep the trucks moving, exchanging empties for the loaded conveyors.[24]

Taking the garments from the bench was an extra non-value-adding step that was eliminated by placing them on the carrier instead. The idea can easily be extended by turning the carrier into a kanban container on wheels; nothing is produced unless an empty carrier is waiting.

SUMMARY: PULL PRODUCTION CONTROL

This chapter focused on various forms of pull and synchronized production control. All are generally preferable to traditional methods in which work is pushed into the production line the way a python swallows an entire pig.

- *Flow lines* are amenable to pull production systems like kanban and drum-buffer-rope. Each workstation can operate at its own pace, so the overall pace should be set by the constraint(s) or capacity-constraining resource(s).
  - DBR has an advantage over kanban in terms of simplicity. Kanban systems require information links between each workstation, but DBR links only the constraint and production starts.
  - *Simplified market pull* (SMP) is like DBR, except that it links the market constraint to production starts.
- *Paced lines* include traditional assembly lines in which a moving conveyor belt dictates the overall system's pace. Their successful operation requires balancing of the operations, so each has roughly the same amount of work.
  - Takt time, which is the available time divided by the amount of required work, should set the overall pace.
  - Music plays a key role in choreographing complex activities like ballets and marching band performances. Henry Ford was similarly successful in choreographing the activities of different moving assembly lines to produce low-inventory flow.

The importance of keeping the product in motion cannot be overemphasized. If cycle time is a performance measurement, work must not be allowed to sit idle in storerooms or warehouses. As Henry Ford pointed out, "Everything has to move in and move out."[25] Chapter 3 discusses variation, as well as its effects on cycle time and inventory, in more detail.

Endnotes

1. Standard and Davis, *Running Today's Factory*, pp. 111–112.
2. Ford, *My Life and Work*, p. 143.
3. Gourley, *Wheels of Time*, p. 00.
4. Ford, *My Life and Work*, p. 82.
5. Hayes, Wheelwright and Clark, *Dynamic Manufacturing*, pp. 50–51.

6. Sorensen, *My Forty Years with Ford*, p. 281.
7. http://www.windjammers.net/shanties.html; the site is no longer online.
8. Parkhurst, *Applied Methods of Scientific Management*, p. 82.
9. The System Company, *How Scientific Management is Applied*, p. 109.
10. The System Company, *How to Get More Out of Your Factory*, p. 27.
11. Hopp and Spearman, *Factory Physics*, p. 361.
12. Smith, *Time Out*, p. 197.
13. Turbide, "A Lean Approach to Lean," p. 0.
14. Hopp and Spearman, *Factory Physics*, pp. 349–360.
15. Ibid., pp. 475–478.
16. Ibid., pp. 475–478.
17. Ibid., p. 640.
18. Ford, *My Life and Work*, p. 82.
19. Hands, *Factory Physics*, pp. 642–645.
20. Schonberger, *Japanese Manufacturing Techniques: Nine Hidden Lessons in Simplicity*, p. 104.
21. Norwood. *Ford: Men and Methods*, p. 19.
22. Sawyer, "Hot off the Skillet."
23. The System Company, *How to Get More Out of Your Factory*, p. 97.
24. Ibid., p. 100.
25. Ford, 0*My Life and Work*, p. 167.

THREE

Variation

When quality practitioners think of variation, they usually think of it relative to part specifications. *Process capability* is, in fact, a ratio of the specification width to a process's standard deviation, as shown in the following equation. A greater ratio is better because it means a smaller fraction of the product will be out of specification due to random chance.

$$\text{Process capability } Cp = \frac{USL - LSL}{6\sigma}$$

where LSL and USL are the specification limits, and σ is the process standard deviation.

Under the Theory of Constraints, however, the variation of interest involves processing time and material transfer time. This is the variation that makes it impossible to run a balanced factory at 100 percent capacity. The following section compares these variation sources to the traditional ones that affect process capability.

VARIATION SOURCES

Quality practitioners are familiar with six traditional variation sources, as reflected in the fishbone or cause-and-effect diagram. These factors reduce the process capability by aggravating variation in the product characteristics:

- Manpower (personnel issues like adequacy of training and worker fatigue)
- Machines (equipment issues like tool wearout and inadequate preventive maintenance)

- Methods (the work instruction; how the job is performed)
- Materials (in general, everything in the bill of materials along with consumables like etching and plating solutions, and machining coolants)
- Measurements (calibration and capability of the gauges and instruments that measure the work and provide feedback data for statistical and other process controls)
- Environment (temperature, humidity, particles in semiconductor manufacturing, and bacteria in food processing)

A quality problem at a given operation in the work stream can generally be traced to one of these sources. A key observation is that the problem source will be very close to the job in question: inadequately trained personnel; equipment deficiencies; poor instructions or methods; defective incoming materials; and so on. The typical operation can, in fact, be viewed according to the SIPOC (Suppliers, Inputs, Processes, Outputs, and Customers) model in Figure 3-1.

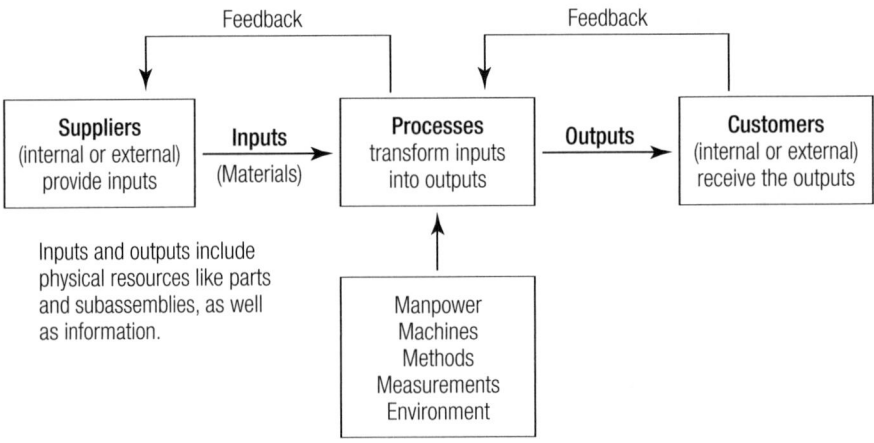

Figure 3-1. SIPOC model.

The SIPOC model suggests that the source of any quality problems is likely to be in direct proximity to the process in question. If the problem lies outside the process itself, it is fairly easy to look to the internal or external suppliers of the inputs. The problem-solving process is conceptually simple, even though its execution may take substantial effort depending on the situation.

Just-in-time production control and its analogues require, however, that the inputs not only meet quality requirements but also arrive no later than they are needed and, ideally, not much sooner. As with quality problems, variation in processing times is usually local to the operation. *Variations in material flow, however, often originate far from the process they affect.* A complex bill of materials aggravates the problem, as shown in the following section.

"Waiting to Match" and Supply-Chain Dependence

Kanban and drum-buffer-rope production control seem relatively simple in textbook situations. In a kanban system, for example, each workstation is in contact with its downstream customers through visual controls like cards or kanban squares. It produces a new item only when it receives a kanban card or sees an empty kanban square. This model creates a visual image of identical workpieces progressing through the factory and having something done to them at each successive station.

The problem becomes more complicated, though, when it is necessary to combine two or more items in a given operation. Waiting to match means that an assembly operation cannot proceed until every necessary item is present, and the problem gets worse as the bill of materials becomes more complex (Figure 3-2).

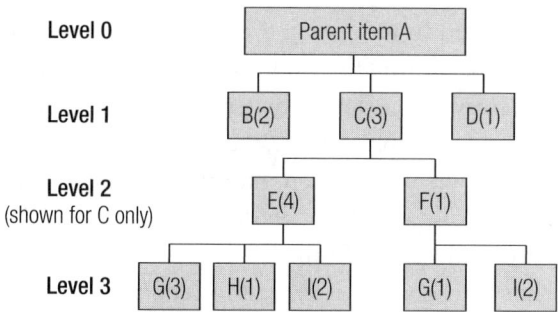

Figure 3-2. Bill of materials.

It is clear from this figure that two units of B, 3 of C, and 1 of D must be at the workstation to manufacture parent item A. *Kitting*, or sending the components to the assembly station as a group, is a fairly standard way to achieve this. Any shortage of E or F, or any variation in their rate of delivery, will, however, affect the availability of C. E depends, in turn, on G, H, and I, and so on.

Figure 3-2 underscores the enormous dependence of smooth production flow on the entire supply chain. A supply chain is "a network that delivers products and services, beginning with raw materials and ending with customers, through engineered flows of information, physical distribution, and cash."[1] If, for example, the bill of material's Level 3 parts come from different suppliers or, even worse, offshore suppliers with long lead times, it is no surprise that the manufacturer may have to keep substantial inventories of these items on hand. It is hardly surprising to find materials managers who must think like the medieval lords who kept enormous inventories of food and other supplies on hand in case someone besieged their castles.

If food supplies were rotated through the castle on a first-in-first-out basis, they were eventually eaten and did not go to waste. In industry, this may be true of undifferentiated materials (like the iron ore and coal that Henry Ford stored outside his steel mill) and standard items like screws and bolts. Hopp and Spearman[2] describe how low-level components may, in fact, be built to stock for later assembly to order. As previously discussed, this may be acceptable if the carrying cost is not high and the inventory cannot serve as a hiding place for defects. In summary:

- Although variation in processing times is local to the operation in question and can probably be addressed through the traditional cause-and-effect diagram, variation in material transfer times can originate far upstream and, possibly, outside the organization.
- The variation is likely to worsen with the complexity and length of the bill of materials, and also with delivery lead times. Offshoring aggravates the delivery lead times.
- Supply-chain management is critical in reducing the variation.

Batching

Process batching refers to the need to collect a given number of parts to fill a batch tool like a heat-treatment furnace or plating bath. *Transfer batching* means collecting enough parts to fill a cart or forklift or, on a larger scale, a truck. Both add to cycle time and variation in material transfer time. (The "Single-Unit Processing" section in Chapter 4 treats the effects of batching in more detail.)

Quality and Reliability Problems

Rework aggravates variation because parts must travel back through the production line and either interrupt or wait for the regular process flow. Standard quality improvement and error prevention techniques are applicable to rework and scrap reduction.

Inventory is not, however, merely a consequence of quality problems; it is also an indirect cause. A basic Japanese manufacturing philosophy is, "Don't take it, don't make it, don't pass it on," with "it" referring to poor quality. That is, no defect should ever get beyond the workstation that produced it, and corrective action should be taken to prevent its recurrence.

This objective is achievable with single-piece flow, as long as there is some way to check the quality of each outgoing piece. If, however, the parts are not checked immediately after production and are sent to a stockroom or warehouse, a considerable amount of defective work may be produced before the trouble is detected. Successive check systems, which have the following key characteristics, support "Don't take it, don't make it, don't pass it on."[3]

- Every piece is inspected.
- The inspector must be impartial (much like a quality auditor).
- If a piece is defective, the operation that produced the defect is notified immediately.
- Corrective action is then taken to eliminate the defect source.

The reference cited here adds that self-check systems are even better. *Self-check systems* improve on successive check systems by immediately checking the quality of each piece as it is produced. Ford Motor Company, for example, used a wide array of snap gauges to make sure each part that left a tool met specifications. Nonconforming work was rejected immediately. The "Inline Quality Control" section in Chapter 5 discusses these ideas in more detail.

Unplanned equipment downtime also adds variation to processing and material transfer times. Chapter 5 discusses the importance of preventive maintenance in avoiding this problem.

This section has shown some possible sources of variation in processing and material transfer times. The following section illustrates the effects of variation. Note, however, that it is a mistake to write the variation off as unavoidable.

EFFECTS OF VARIATION: THE MATCHSTICKS-AND-DICE EXERCISE

The matchsticks-and-dice simulation, as adapted from Goldratt and Cox's *The Goal*, works as follows. Each workstation can process one die roll's worth of WIP, if it is available at the workstation. The procedure is to roll one die and transfer the lesser of the die roll and the amount of work available. For example, if the die roll is a four but only three units are waiting to be processed, the workstation can ship only three. Because the average die roll is 3.5, we expect the simulated production line to ship an average of 3.5 units per turn, but we will soon discover that actual production is far less. This is because of a hurry-up-and-wait effect in which high die rolls are wasted when no work is available. Because each workstation is a constraint (given that the factory is completely balanced) and time lost at a constraint is lost forever, every such incident means a permanent loss of throughput.

Table 3-1 shows the setup for a visual basic simulation of the matchsticks-and-dice experiment. Each of four workstations begins with four units (a little more than the 3.5 that each is expected to process in the first turn). The purpose of this starting inventory is to reduce the time it will take for the system to achieve a steady state, if one can ever be achieved. Otherwise it would take several turns before the downstream workstations could produce anything at all.

Table 3-1. Beginning setup for matchsticks-and-dice simulation.

Workstation	Release	1	2	3	4	Finished
Inventory		4	4	4	4	
Die roll						
Pass to next						
Change						

Note how each workstation passes the lesser of its die roll (its production capability for the simulation turn) and its work-in-process inventory.

VARIATION

Table 3-1a. First die roll for matchsticks-and-dice simulation.

Workstation	Release	1	2	3	4	Finished
Inventory		4	4	4	4	
Die roll	1	1	1	6	6	
Pass to next	1	1	1	4	4	
Change		0	0	-3	0	

Although workstations three and four rolled sixes, they can process only the four parts they have. Notice especially how the four workstations rolled, on average, the expected production rate of 3.5 units per turn. Examination of the "pass to next" row shows, however, that the workstations moved only ten pieces for an average of 2.5.

Since the die rolls will always average 3.5, it is clear that the four wasted die spots (opportunities to make product) will never be recovered. Time lost at a constraint is lost forever, and every operation in a balanced process is a constraint. This is a key lesson of the matchsticks-and-dice exercise.

The bottom row shows the change in WIP inventory at each workstation. Since workstation four shipped four units while receiving one unit from workstation two, its WIP declines by two units. The changes for the other workstations are zero, because they received as many units as they shipped. The respective inventories at the start of the second turn will be four, four, one, and four respectively, as shown in Table 3-1b. Note how the "finished" column also keeps track of the number of turns and the average production rate.

Table 3-1b. WIP after transfer of completed units, first simulation turn.

Workstation	Release	1	2	3	4	Finished
Inventory		4	4	1	4	4
Die roll						N= 1
Pass to next						Avg= 4.
Change						

Table 3-2 shows the results of the second set of die rolls. The production die rolls are actually very good. Their average, (6 + 3 + 2 + 6) / 4 is 4.25, well above the expected 3.5. Two die spots are, however, wasted at workstation one, because it has only four units available to process. Another spot is wasted at workstation three. Although we begin this exercise with the natural expectation that high die rolls will offset low ones, the wasted die spots from the high rolls are irretrievably lost, so this system will never achieve the expected average of 3.5 units a turn.

The question now arises as to whether the situation will eventually stabilize with so much WIP at each workstation that future die rolls will never be wasted. The visual basic program can simulate up to 7,000 turns so the

Table 3-2. Second set of die rolls.

Workstation	Release	1	2	3	4	Finished
Inventory		4	4	1	4	4
Die roll	5	6	3	2	6	N= 1
Pass to next	5	4	3	1	4	Avg= 4.
Change		1	1	2	-3	

system's steady state, if there is one, can be achieved and visualized. Table 3-3 shows two long-term simulations, by which time a steady state might be expected. The bar chart at the right of the first screenshot is actually a miniature of the one that appears on the program, and its purpose is to show the bubble or ripple effect that seems to accompany this simulation. In Table 3-3, workstations one, three, and four have so much work waiting that all die spots will be utilized. There is, however, a five-out-of-six chance (that

Table 3-3. Simulation after 823 die rolls.

Workstation	Release	1	2	3	4	Finished
Inventory		45	1	57	30	2741
Die roll						N= 823
Pass to next						Avg= 3.33
Change						

is, any die roll higher than one) that die spots will go to waste at workstation two. In Table 3-3a, which gives a simulation's status after 2,149 die rolls, there is a 50 percent chance that die spots will go to waste at workstation four even though there are 193 units of WIP among the four workstations!

Table 3-3a. Simulation after 2,149 die rolls.

Workstation	Release	1	2	3	4	Finished
Inventory		22	49	119	3	7316
Die roll						N= 2149
Pass to next						Avg= 3.4
Change						

VARIATION

Table 3-4 shows a simulation after 7,001 turns, at which time we would expect 24,503.5 finished units. There are only 24,070. The shortfall is 433.5, and there are, in fact, 483 units stuck in the system as inventory.

Table 3-4. Long-term steady state status.

Workstation	Release	1	2	3	4	Finished
Inventory		221	120	47	95	24070
Die roll					N= 7001	
Pass to next					Avg= 3.44	
Change						

Figure 3-3 plots both the average production rate (finished units divided by turns, the uppermost of the two lines) and total WIP inventory with time. It suggests that the expected production rate of 3.5 per turn will never be achieved, which is intuitively apparent given the waste of high die rolls at a workstation with only a couple of units waiting. It also makes the counterintuitive suggestion that although the total WIP inventory may fluctuate, it seems to trend upward forever.

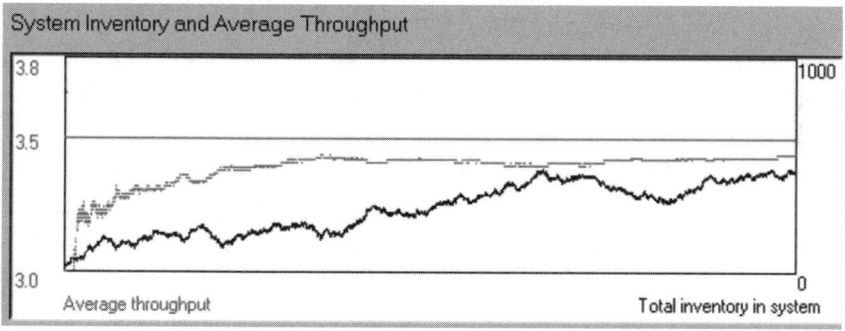

Figure 3-3. Performance graph.

Here is a summary of the underlying problem. Because of variation in processing and transfer times from upstream workstations:

- Work may not be available when a workstation has capacity. Because each workstation is a constraint (balanced factory, all stations have equal capacity), the wasted capacity is lost forever.

- Too many units may arrive from the upstream workstation, which increases inventory and waiting time.

Factories are not, incidentally, the only setting in which variation keeps throughput below capacity. Sharon Begley describes "phantom traffic jams" that are apparently triggered by variations in traffic flow.[4] They are often exacerbated by weaving as drivers seek faster lanes—much as production managers try to expedite priority jobs. "That invites yet more cars to change lanes, propagating a wave of stop-and-go traffic that cuts the number of cars in a stretch of road by about 10 percent," a phenomenon that can cause a five-mile backup during rush hour. Braking to accommodate merging traffic forces trailing vehicles to do the same, and the effect can propagate backward for miles.

Even more frightening is Begley's statement, "Like water that suddenly freezes, flowing traffic can spontaneously seize up, beginning at a single point of crystallization (the idiots who braked to rubberneck) and causing a wave of high density to spread backward." It is easy to envision the same thing happening in a factory, especially one that allows work to be pushed into the line regardless of available capacity.

It is possible to envision rush-hour traffic moving very quickly and smoothly if every car's speed could be made equal. This might be achievable in the future if sufficiently reliable vehicle control systems are developed. A driver would simply indicate his or her destination, and then a traffic control computer would set the vehicle's speed while perhaps even keeping it in its lane—thus freeing the driver to read or use a laptop computer. Needless to say, the system would have to be absolutely reliable so that it wouldn't cause massive traffic accidents.

If such a system were developed, the road could accommodate more vehicles. Human drivers need at least a second to react to braking by the vehicle ahead of them, and cushions of two seconds or greater are recommended. The safety cushion is, in fact, very similar to inventory whose purpose is to protect the system—or in this case the drivers—from variation. The limitations of the human nervous system mandate a relatively large safety cushion. A computerized traffic control system could, in contrast, leave a fraction of a second between vehicles without risking a rear-end collision, because it would remove the variation from the highway.

VARIATION

Although such an arrangement for highway traffic might be a couple of decades in the future, the same concepts are applicable to today's factories. Recall that Henry Ford succeeded in pacing an assembly line without the aid of any computers.

This section has used the matchsticks-and-dice simulation to illustrate the effects of variation on a balanced factory that is operating at 100 percent capacity. The following section examines performance characteristics of a single-server queue and introduces the equation that connects the variation concept to Henry Ford's success in running a balanced factory at close to full capacity.

EFFECT OF VARIANCE: PERFORMANCE CHARACTERISTICS OF SINGLE-SERVER QUEUE

Hopp and Spearman[5] show how to calculate the work-in-process and cycle time of a single-server queue, given server utilization u. Let the system contain n units in queue.

- The only possible changes to the system are an increment $(n + 1)$ when a new unit arrives or a decrement $(n - 1)$ when a unit is processed.
- The server is idle only when there is no WIP in the system, so $p_0 = 1 - u$ (where p_n is the chance of having n units in queue).
- The steady-state rate at which the system moves from state $(n - 1)$ to n is $p_{n-1}r_a$, where p_n-1 is the chance of having $n - 1$ units in queue and r_a is the arrival rate. The arrival rate (items per unit time) is the reciprocal of the mean arrival time.
- The steady-state rate at which the system moves from state n to $(n - 1)$ is $p_n r_e$, where p_n is the chance of having n units in queue and r_e is the processing rate. The processing rate (items per unit time) is the reciprocal of the mean processing time.

The existence of a steady state requires that $p_{n-1}r_a = p_n r_e$; that is, the rates of increase and decrease must be equal if the system is to remain in equilibrium. Now:

$$u = \frac{r_a}{r_e} \Rightarrow p_n = p_{n-1}u$$

Since $p_0 = 1 - u$, $p_1 = u(1 - u)$ and $p_n = u^n(1 - u)$

$$WIP = \sum_{n=1}^{\infty} np_n = \sum_{n=0}^{\infty} nu^n (1-u) = (1-u)\sum_{n=1}^{\infty} nu^n$$

The first term of the series is zero and therefore

$$WIP = (1-u) \sum_{n=1}^{\infty} nu^n = (1-u)u \sum_{n=1}^{\infty} nu^{n-1}$$

There is an infinite series that states:

If $u < 1$ then $\dfrac{1}{(1 \mp u)^2} = 1 \mp 2u + 3u^2 \mp 4u^3 + \cdots$ so $\dfrac{1}{(1-u)^2} = \sum_{n=1}^{\infty} nu^{n-1}$

$$WIP = (1-u)u \sum_{n=1}^{\infty} nu^{n-1} = \frac{(1-u)u}{(1-u)^2} = \frac{u}{1-u}$$

Furthermore, Little's Law states that cycle time equals WIP divided by throughput. This leads to the following equation set:[6]

- Work in process $WIP = \dfrac{u}{1-u}$

- Cycle time $CT = \dfrac{WIP}{r_a} = \dfrac{\frac{u}{1-u}}{\frac{u}{t_e}} = \dfrac{t_e}{1-u}$ since, in steady state, throughput = r_a

- Cycle time in queue $CTQ = CT - t_e = \dfrac{t_e}{1-u} - t_e = \dfrac{t_e u}{1-u}$

  since time in queue = total cycle time minus time spent being processed

- WIP in queue $WIP\ q = r_a CTQ = \dfrac{u}{t_w} \dfrac{t_e u}{1-u} = \dfrac{u^2}{1-u}$

Standard and Davis cite the following equation for cycle time in queue when interarrival times follow the exponential distribution.[7] Hopp and Spearman refer to this equation as the Kingman equation (after its first investigator) or VUT (variability-utilization-time) equation. They state that it is exact for the M/M/1 queue and also for the G/G/1 queue, where the 1 means "one server." (M stands for the exponential [memoryless] distribution and G for nonexponential distributions.[8] Avrill M. Law and W. David Kelton use the exponential distribution to model interarrival times and service [or processing] times in a single-server or M/M/1 queue.)[9] The Kingman equation:

$$CT_q = \left(\frac{c_a^2 + c_e^2}{2}\right)\left(\frac{u}{1-u}\right) t_e \text{ where}$$

CT_q = cycle time in queue, waiting for the workstation

VARIATION

$c_a = \dfrac{\sigma_a}{t_a}$ = coefficient of variation, arrivals at the workstation

$c_e = \dfrac{\sigma_a}{t_e}$ = coefficient of variation, effective processing time

$u = \dfrac{t_a}{t_e}$ = utilization

t_a = average time between arrivals at the workstation

t_e = effective processing time

The practical implications of this equation lie in the factors $\left(\dfrac{c_a^2 + c_e^2}{2}\right)$ and $\left(\dfrac{u}{1-u}\right)$. The second factor says that cycle time in queue (and, hence, inventory) will indeed approach infinity as utilization approaches 100 percent. The first shows, however, that elimination of all variation in material transfer and processing times will reduce cycle time and inventory to zero—and there is extensive evidence that this was Henry Ford's production strategy.

As noted above, Hopp and Spearman[10] state that the Kingman equation is exact for the M/M/1 queue and also for the G/G/1 queue, where the 1 means "one server." The former assertion is easy to prove, noting that an exponential distribution's mean (θ) and standard deviation are identical. This makes the coefficient of variation 1 for any exponential distribution, whereupon the Kingman equation reduces to $CTQ = \dfrac{t_e u}{1-u}$.

The exponential distribution is, in fact, a standard model for the times between random arrivals. Its counterpart, the poisson distribution, is the model for random arrivals per unit time. This makes the exponential distribution the most logical model for modeling the times between arrivals of customers at a server or parts at a workstation.

It is intuitively reasonable to use the exponential distribution for processing times as well, since there is usually a certain minimum time in which the task can be performed. As an example, a stamping operation's speed may be limited by the speed at which the press opens and closes. The exponential distribution can account for this minimum through a *threshold parameter*, which is the same thing as the guarantee time in reliability applications.

$$f(t) = \frac{1}{\theta} \exp\left(\frac{-(t-\delta)}{\theta}\right)$$

where δ is the threshold parameter (t cannot be less than δ), and θ is the shape parameter. (In reliability applications, θ is also known as the mean time between failure or characteristic life.) $\theta + \delta$ is the average of a pure exponential distribution with no threshold parameter, so $\theta + \delta$ is the mean of an exponential distribution with a threshold parameter.

Figure 3-4 shows an exponential distribution. The task cannot be completed in less than 10 units of time ($\delta = 10$), and random variation (we will question its true randomness later) usually causes it to take more. Since $\theta = 1$, the average task time is 11 units of time.

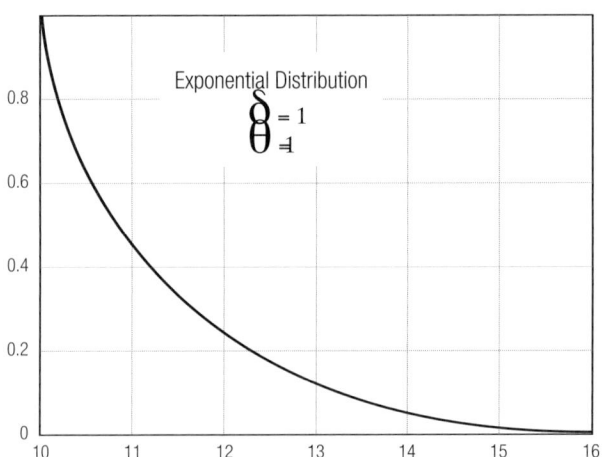

Figure 3-4. Exponential distribution with threshold parameter.

SUMMARY: VARIATION

This chapter illustrates some typical sources of variation in processing and material transfer times. Batching, whether for processing or transfer, tends to aggravate this variation while increasing cycle times as well. Waiting-to-match is a problem in assembly organizations, and it is aggravated by complicated bills of material. Kitting is a traditional way of dealing with this problem. Quality problems that result in rework and scrap, along with reliability problems that cause equipment downtime, also add to this variation.

The matchsticks-and-dice exercise from *The Goal* shows that the effect of this variation is to starve the constraints of a perfectly balanced system so that it always falls short of its theoretical capacity. Such starvation can even take place in the midst of plenty; one station can have a huge backlog of inventory while another waits for work. This underscores the lesson that variation in processing and material transfer time is the mortal enemy of smoothly flowing production.

The problem with any random distribution model, whether it is exponential, normal, or the uniform integer range from a die roll, is that it reinforces the self-limiting paradigm that the factory is at the mercy of variation, and nothing can be done about it. The matchsticks-and-dice simulation is excellent for teaching the *effects* of variation, but it becomes dangerous if people take away the lesson that they have to live with the variation. Chapter 4 discusses ways to reduce the variation so the factory can roll a six, or close to it, every time.

Endnotes

1. Walker, "Supply Chain Management," APICS-NEPA meeting, Pittston, PA, March 14, 2001.
2. Hopp and Spearman, *Factory Physics*, p. 322.
3. Shingo, *The Sayings of Shigeo Shingo*, p. 76.
4. Begley, "How Brief Drop in Cars Can Trigger Tie-Ups, and Other Traffic Tales," p. 0.
5. Hopp and Spearman, *Factory Physics*, pp. 267–269.
6. Ibid., p. 269.
7. Standard and Davis, *Running Today's Factory*, p. 234.
8. Hopp and Spearman, *Factory Physics*, p. 270.
9. Law and Kelton, *Simulation, Modeling, and Analysis*, pp. 15–16.
10. Hopp and Spearman, *Factory Physics*, p. 270.

FOUR

Variation Reduction

*The fault, dear Brutus, is not in our stars
But in ourselves, that we are underlings*[1]

This chapter shows why variation in production and material transfer times is not in our stars either, and that something can be done about it. There will be some similarity between this chapter and the one that follows because most productivity improvement techniques and principles also have the tendency to reduce variation. This chapter, therefore, focuses on those whose primary effect is to suppress variation in material transfer times, while Chapter 5 addresses more general lean manufacturing principles.

A common complaint against Henry Ford's moving assembly line was its purported tendency to dehumanize the worker through mindless and repetitive work. As Ford himself wrote in *My Life and Work*, "The man who places a part does not fasten it—the part may not be fully in place until after several operations later. The man who puts in a bolt does not tighten it."[2] Charlie Chaplin took this job design concept to comic extreme in *Modern Times*, in which he mindlessly tightened bolts on an assembly line, and then sought to apply his wrenches to every bolt-like object in sight.

The truth is, however, that even self-directed groups of workers subdivided tasks because this made the work easier. The following section explains how subdivision of labor reduces variation while making a job easier.

SUBDIVISION OF LABOR

Rudyard Kipling's *Captains Courageous* describes a fish-cleaning operation in which a fishing boat's crew created its own assembly (or rather disassembly) line. The subsequent discussion shows how this job design reduced cycle time and variation, but was still far from optimum. Try to identify a glaring example of waste in the following job description (an answer will be given afterward).

Penn and Manuel stood knee-deep among cod in the pen, flourishing drawn knives. Long Jack, a basket at his feet and mittens on his hands, faced Uncle Salters at the table, and Harvey stared at the pitchfork and the tub.

"Hi!" shouted Manuel, stooping to the fish, and bringing one up with a finger under its gill and a finger in its eye. He laid it on the edge of the pen; the knife-blade glimmered with a sound of tearing, and the fish, slit from throat to vent, with a nick on either side of the neck, dropped at Long Jack's feet.

"Hi!" said Long Jack, with a scoop of his mittened hand. The cod's liver dropped in the basket. Another wrench and scoop sent the head and offal flying, and the empty fish slid across to Uncle Salters, who snorted fiercely. There was another sound of tearing, the backbone flew over the bulwarks, and the fish, headless, gutted, and open, splashed in the tub, sending the salt water into Harvey's astonished mouth. After the first yell, the men were silent. The cod moved along as though they were alive, and long ere Harvey had ceased wondering at the miraculous dexterity of it all, his tub was full.

The following Ford job design principles are obvious:

- Three men subdivided the "whole task" of cleaning a fish among themselves, which allowed them to process the fish far more quickly than by working individually.
- The product is in continuous motion: "The cod moved along as though they were alive"—and this simple sentence should provide a mental picture of the ideal condition for any production operation. The workers are close enough to one another so that they can pass (or throw) the fish to the next person, without anyone having to walk.

The job design failed, however, to incorporate yet another Ford principle: No job should require a worker to bend over, especially not for repetitive lifting. The work should be delivered roughly at waist level, but "Penn and Manuel stood *knee-deep among cod* in the pen . . . 'Hi!' shouted Manuel, *stooping to the fish*, and *bringing one up* with a finger under its gill and a finger in its eye." A redesign of the fish pen would doubtlessly have saved them considerable physical effort while allowing them to process even more fish and thereby earn more money, while ending the work day with fewer aching muscles.

If the reader missed this, the author also admits to overlooking it when citing this example in *Henry Ford's Lean Vision* (2002), even though the passage was used to illustrate the variation-reducing aspects of subdivision of labor. This oversight underscores the value of gemba leadership (that is, going to the value-adding workplace and seeing for oneself as opposed to trying to manage from an office) and also the benefit of videotaping the job. It is easy to overlook the waste in this job from a written description, but it would have been obvious on sight to Henry Ford or Frank Gilbreth (who prescribed waist-level delivery of bricks to masons) the instant they saw a man bend over to pick up a fish.

It is very instructive to look at the fish-cleaning process from a mathematical perspective. It is doubtful that the fishermen knew what standard deviation and variance are, but they certainly figured out their effects on their own. Henry Ford never showed any understanding of industrial statistics (which were in their infancy when he was building his business), but everything he did focused on reducing variation. The following equation shows how to compute the standard deviation of a sum of random numbers, like task times in a multistep process.

$$\sigma_{TOTAL} = \sqrt{\sum_{i=1}^{n} \sigma_i^2}$$

Table 4-1 applies this equation to the fish-cleaning process as it might be performed by three people doing the entire job as individuals.

Table 4-1. Fish-cleaning process.

Task	Time	Standard deviation
(1) Pick up knife	t_1	σ_1
(2) Cut fish open and notch behind head	t_2	σ_2
(3) Put knife down	t_3	σ_3
(4) Remove liver and drop in basket	t_4	σ_4
(5) Pull off and discard head and entrails	t_5	σ_5
(6) Remove the fish's spine and put the fish into the bucket of salt water	t_6	σ_6

The total cycle time and standard deviation are, therefore,

$$T = \sum_{i=1}^{6} t_i \text{ and } \sigma_{TOTAL} = \sqrt{\sum_{i=1}^{6} \sigma_i^2}.$$

Table 4-2 shows the effect of subdividing the same task among three workers. This immediately eliminates the waste motion of picking up and putting down the knife for each fish.

Table 4-2. Subdivided fish-cleaning process.

Task	Time	Standard deviation
First worker: (2) Cut fish open and notch behind head	t_2	σ_2
Second worker: (4) Remove liver and drop in basket	t_4	σ_4
Second worker: (5) Pull off and discard head and entrails	t_5	σ_5
Third worker: (6) Remove the fish's spine and put the fish into the bucket of salt water	t_6	σ_6

It is immediately obvious that, no matter what the times and standard deviations are for each task, $t_2 + t_4 + t_5 + t_6 + < \sum_{i=1}^{6} t_i$ and $\sigma_2^2 + \sigma_4^2 + \sigma_5^2 + \sigma_6^2 + < \sum_{i=1}^{6} \sigma_i^2$. Now Ford's statement, "The man who places a part does not fasten it.... The man who puts in a bolt does not tighten it," becomes far more meaningful. Picking up and putting down a tool for each unit is non-value-adding waste and it adds variation to processing time.

There is, however, a caveat that goes with subdivision of labor. Subdividing labor increases the productivity of each worker by eliminating non-value-adding activities, but unless the tasks are balanced, it creates the waste of waiting. In a balanced production line, everyone has the same amount of work. No one has so much to do that he or she gets behind, but no one has to wait because he or she has too little to do. As Ford observed, "A man must not be hurried in his work—he must have every second necessary but not a single unnecessary second."[3] Each task should, therefore, require the same amount of time.

In the fish-cleaning example from *Captains Courageous*, Manuel may have had too much to do because he had to bend over for each fish. This means that Long Jack and Uncle Salters probably lost time waiting for the fish. On the other hand, if the fish pen was redesigned to eliminate the need for stooping, Manuel might have passed the fish to Long Jack too quickly. In Theory of Constraints language, elevating a constraint to the point where it is no longer a constraint must create a new one. In this case, Long Jack would have had to set the pace for the entire operation, perhaps with the equivalent of a two-fish replenishment system. If two fish were waiting at Long Jack's table, Manuel would have stopped working; he would slit and throw fish only when the "kanban square" contained one or zero.

It is, in fact, usually easier to manage unbalanced lines than balanced ones, because the constraint sets the pace for all the other operations. Under drum-buffer-rope production control, the constraint needs to communicate only with the first operation (or production releases) while letting everything in between work at its own pace. To assess whether such a situation is desirable, simply ask whether the imbalance wastes the time of things or people. Although people should not produce unusable inventory, they cannot be paid high wages for waiting for work any more than they can be paid (as Henry Ford pointed out) for walking to get parts or tools. This idea suggests that people's work should be balanced even if the work done by the equipment is not. Japanese production lines are often designed to allow a worker to operate more than one machine at a time.

The skillet (described in Chapter 2), which moves the worker with the job, also keeps the worker's time fully and productively occupied. It neither wastes his time by forcing him to wait for work nor risks overloading him with more than he can handle. The drawback is that it may require non-value-adding work like picking up and putting down tools, as previously described. Alternatively, suppose the fish-cleaning process from *Captains Courageous* was, in fact, balanced. The men might then have worked in time to rhythmic music provided by a shantyman, as previously described.

Job redesign to reduce variation and non-value-adding motion is very helpful in "rolling a six every time" in a matchsticks-and-dice context, but intelligent production planning can be even more important. *Value stream analysis* helps identify non-value-adding activities, which (as illustrated in the fish-cleaning example) add both cycle time and variation to the process.

VALUE STREAM ANALYSIS

Value stream analysis is synergistic with process flowcharting, one of the seven traditional problem-solving tools. (The others are the tally sheet or check sheet, histogram, Pareto chart, statistical process control chart, scatter diagram, and fishbone or cause-and-effect diagram.) John Hradesky[4] recommends using the following symbols for graphic representations of a value stream:

- Circle = operation (processing)
- Square = 100 percent inspection
- Arrow = transportation
- Inverted triangle = storage

A diamond is often used to indicate decisions in computer flowcharts, and the same symbol might be used in a process flowchart to show what happens to a product, depending on its status. As an example, nonconforming units may be sent for rework. Note, however, that only operations designated by a circle (processing) add value. The others should generally be regarded as necessary (that is, value-assisting) or as unnecessary evils that add both cycle time and time variation to the process.

Even a circle in the flowchart does not guarantee that the activity it represents is value-adding. Axle manufacturing at the Ford Motor Company during the 1910s or early 1920s involved a heat treatment step followed by a straightening operation to correct distortions from the heat treatment. Because both activities transform the product, they would have been designated by circles, and factory personnel might have assumed that they were value-adding. The truth is that the straightening operation was 100 percent rework because it corrected a problem that was inherent in the preceding operation. The company eventually eliminated the straightening operation by introducing a centrifugal hardening machine that ensured that the shafts cooled evenly when they came out of heat treatment.[5]

Workflow Analysis

The Automotive Industry Action Group (AIAG) describes *workflow analysis* as an easily used form of value analysis.[6] Tasks are defined and then arranged in proper sequence (this is consistent with basic process flowcharting). Each task or activity is then categorized as one of the following:

- Operation: an activity that transforms the part (that is, the value-adding "Bang!")
- Transportation (non-value-adding)

- Inspection (value-assisting)
- Delay or temporary storage (non-value-adding)
- Storage (non-value-adding)
- Decision (value-assisting)

A worksheet provides a column for recording the time for each activity; another column is used for identifying activities that are bottlenecks or waste. Two transportation arrows in series are evidence of potential waste. The example provided in the AIAG reference includes the steps, "Move to receiving inspection" and "move to inspection bench." This suggests that the pieces must be handled more than once. In single-unit flow, the parts would move directly to the inspection bench one at a time.

Interestingly enough, the example defines "receive material" as an operation even though it does *not* transform the part's form, fit, or function. It actually sounds like non-value-adding handling, and it is the bottleneck in the bargain. Of the twelve listed steps, only one (assemble the part onto the engine) is actually value-adding; the two inspections are value-assisting. Receiving might be better defined as an inspection if the items are counted and/or verified as to part number.

Shingo[7] cites Frank Gilbreth's assertion that a process can be broken down into four (not six) kinds of activities:

- Processing
- Inspection
- Transportation
- Delays

Consolidation is attractive because fewer categories make the analysis simpler. The six AIAG categories can presumably be consolidated as follows:

- Inspections and decisions, which are likely to be based on inspections, are redefined as inspections (value-assisting).
- Delay, temporary storage, and storage are all classified as delays (non-value-adding).

Gilbreth's own classification of human motion into eighteen categories or *therbligs* (a rearrangement of his own name), suggests that a five-category system may actually be best—not as a compromise between AIAG's six and

Gilbreth's four, but as the minimum number of classifications necessary to force waste to become visible. Eight of the eighteen therbligs involve grasping and releasing, finding, selecting, and positioning.[8] These are all forms of handling or manipulation that, while they may be necessary, do not transform the part. This is important because handling can otherwise be viewed as an aspect of processing. If, for example, clamping and unclamping are lumped with "processing," they are assumed to be value-adding and no further thought is given to them. If such activities are identified and classified separately, however, people think of ways to consolidate or eliminate them. Unitary machines, which are discussed in the "Work Cells and Unitary Machines" section later in this chapter, perform several value-adding transformations for one grasp-and-release cycle. The five-category system is, therefore:

- Transform (the value-adding "Bang")
- Handle (value-assisting or non-value-adding)
- Transport (non-value-adding)
- Inspect (value-assisting)
- Delay (non-value-adding)

Value stream analysis can be performed in an office or a team meeting room with basic process flowchart information. Workflow analysis requires activity times, but these could conceivably be "should-be" times from industrial engineering activities. In many cases, it might be far more instructive to track actual jobs through the production line and account for exactly how they spend their time.

Cycle Time Accounting

The traditional view of "time accounting" is the apportionment of labor or professional time to individual activities or clients, which can then be charged for them. This section introduces the widely overlooked but quite valuable concept of *accounting for the time that parts spend in the factory*.

If a particular routing requires a total of 1.5 hours of work, why does the job remain days or even weeks in the factory? A couple of decades ago, production control systems gated jobs into and out of different operations, but this provided little detail as to what was actually done with the time in question. Modern bar coding and radio frequency identification (RFID)

VARIATION REDUCTION

technology should make it practical to track every job, or at least sample jobs, through the factory and discover exactly what occurs during the time a job is on site.

To perform this exercise, it is necessary to use the production tracking system to effectively attach a stopwatch to each job. The watch starts the instant the job begins an activity (whether value-adding or non-value-adding) and stops the instant the activity ends. At that point, the clock must *immediately* start for the next activity. There is no need to invent a new quality tool or chart to display this information; the traditional Gantt chart is quite suitable, as shown in Figure 4-1. In this case, however, the Gantt chart does not display *planned* activity times but *realized* activity times. Its purpose is to force the wasted time to become visible and to pinpoint the waste's location.

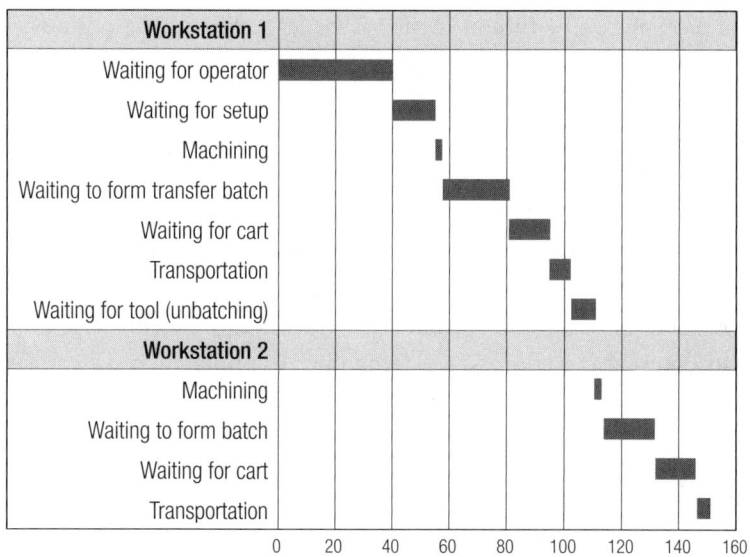

Figure 4-1. Gantt chart display of realized activity times.

The bars can be color-coded—for example, green for value-adding activities like machining, yellow for non-value-adding transportation in which the work is actually moving from one operation to another, and red for waiting. The familiar Pareto chart, which sorts the bars from largest to smallest, is also suitable for displaying this information. Either instrument facilitates the identification of the biggest time-wasters.

In the preceding example, "waiting for operator" may contribute to downtime because of a meal break or shift change. This is not necessarily a problem, except at a capacity-constraining resource that must never be left idle for lack of work or personnel. It may be a one-in-ten occurrence, so it is probably desirable to average the results for many jobs that use the routing in question. On the other hand, "waiting to form transfer batch" and "waiting for the tool" (because the parts arrive in a batch) indicate a non-value-adding contributor to cycle time and possibly flow variation.

Once the sources of non-value-adding cycle time are identified, traditional problem-solving techniques can be used to address them. The importance of identifying and removing the problem's root cause cannot be overemphasized. As General Curtis LeMay aptly observed, swatting flies (fixing the problem every time it recurs) is far less effective than removing the manure pile that is attracting the flies. The "Five Whys" approach is one way to achieve this.

"Five Whys" Technique

Taiichi Ohno[9] and AIAG[10] advocate the *Five Whys* process. This technique is also known as the "question to the void," which means asking "why" until no further progress is possible. This is the point at which the problem's root cause is likely to be found. Antoine de Saint Exupéry's *The Little Prince* provides an entertaining and easy-to-remember example:

> **Q:** "Why do you drink?"
> **A:** "To forget."
> **Q:** "To forget what?"
> **A:** "That I am ashamed."
> **Q:** "Why are you ashamed?"
> **A:** "I am ashamed of being a drunkard!"

The Five Whys technique is normally used to diagnose quality problems, but there is no reason why it cannot be applied to non-value-adding cycle time as well:

> **Q:** Why did the job wait forty minutes before going into the machine?
> **A:** No operator was available.

VARIATION REDUCTION

Q: Why was no operator available?
A: It was the workers' lunch break.
Q: Why were all the operators at lunch at the same time?
A: They have always done it that way.

The clear and simple solution to this problem is to stagger operator breaks so that there is always someone available to run the machine. On the other hand, if the answer to the second question is, "The only operator who knows how to run the machine was at lunch," cross-training is the obvious solution.

A similar Five Whys example can be used to detect and reduce the amount of waiting associated with placement or movement of parts, materials, or tools:

Q: Why did the parts sit at the workstation for 25 minutes after the operator finished them?
A: It took that long to accumulate a transfer batch.
Q: Why must we accumulate a transfer batch?
A: It is not economical for a hand cart or forklift to move smaller batches.
Q: Why must parts be moved by hand cart or forklift? (Also, "Can the carts or forklifts be shared by smaller transfer batches from different workstations?")
A: The tools are not close enough for rollways or conveyor belts.
Q: Why are the tools not close enough for rollways or conveyor belts?

At this point, the answer may be that the tools are "monuments" that cannot be moved because they are on special foundations or require special utility connections. If, however, the only answer is that all the drills are in the drilling department, it may be possible to rearrange the tools into work cells.

The key point of the Five Whys is that the investigator does not take the situation for granted or accept explanations like, "That is how we have always done it" or "That is the way it is." These explanations are really a self-limiting form of inertia. Henry Ford, for example, was very quick to ask *why* the farmer keeps carrying heavy buckets of water instead of taking the practice for granted. If the farmer answers, "There is no pipe to carry the water," he has pretty much found his own solution to the problem.

SINGLE-UNIT PROCESSING

Lean manufacturing doctrine strives for single-unit processing because it promotes flow, and this section shows its importance in suppressing flow variation as well. The plug flow reactor (PFR) introduced in Chapter 2 illustrates the advantages of continuous flow in manufacturing. Batch processes increase cycle time and variation, and they are less amenable to statistical process control and engineering process control. The calculation of statistical process capability indices for batch processes, which is not even treated in most SPC books, is more difficult and less reliable than that used for single-unit operations. The following sections discuss the kind of batching that takes place in factories, examine its effects, and introduce some alternatives.

Process Batching and Transfer Batching

Hopp and Spearman point out that there are actually two kinds of batches: process batches and transfer batches.[11] As the name implies, *process batching* refers to aggregation of units for processing. Process batches can be parallel, which means that the workstation processes all the units simultaneously. This is typical of equipment like furnaces and plating baths, and it is the *parallel batch* that complicates SPC, as shown in "Parallel Process Batch Processes are Harder to Control" later in this chapter.

A *serial batch* consists of a run of parts that are processed individually. Serial batching increases cycle times if the entire run must be completed before the work proceeds to the next operation but not if the parts can be transferred individually. *Lot splitting* is a frequent compromise between waiting for completion of the entire run and shipment of individual pieces. Serial batching is encouraged if shop supervisors are reluctant to "break a setup," and runs can be shortened through use of single-minute exchange of die (SMED).

Transfer batching refers to the need to accumulate a certain number of pieces before sending them to the next workstation. It is, in fact, the in-house analogue of the less-than-load (LTL) issue in trucking. Frequent trips by half-empty trucks or half-empty forklifts and carts within the factory add to the per-unit handling cost, but waiting for full loads increases cycle time and variation due to batching. Transfer batching is not a problem in a moving assembly line, where the transfer batch is a single unit.

VARIATION REDUCTION

Shingo[12] describes the related concepts of *process delay* and lot delay. A process delay occurs when items must wait for a tool because the tool itself is busy with another lot. A *lot delay* means that, while each item in a lot is being processed, the others must wait. The unfinished units must wait for the tool to become available, while the completed ones must wait for the others to go through the tool. Shingo adds that process delays can be addressed by synchronizing the processes (for example, takt time and pacing). The only way to deal with lot delays is through single-unit flow in processing, transportation, and inspection activities.

The following section illustrates some of the disadvantages of batching.

Batching Increases Cycle Time and Variation

Serial batches do not necessarily increase cycle time and variation, but parallel process batching and transfer batching do. They require parts to wait at the batch tool until a full load accumulates (Figure 4-2), and/or wait at a single-unit tool when they emerge from the batch tool.

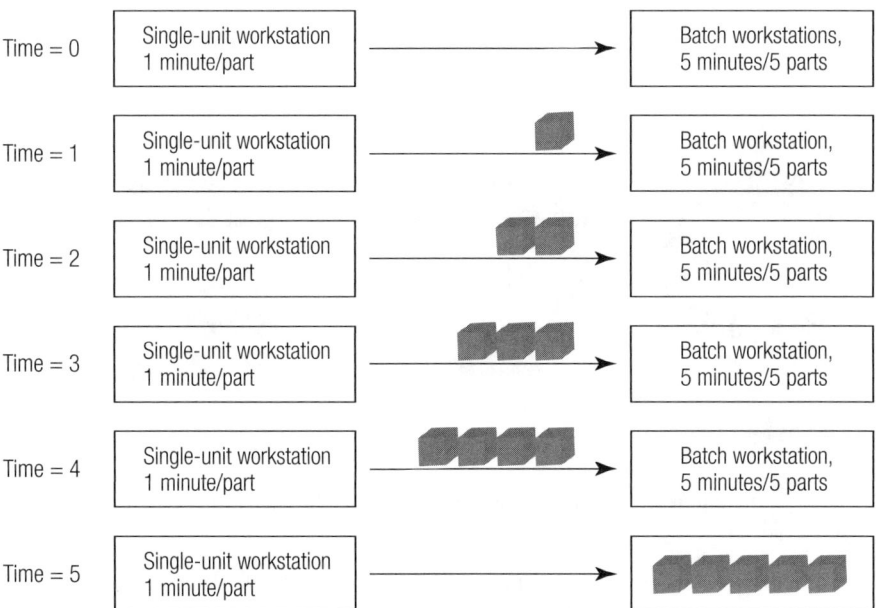

Figure 4-2. Waiting for accumulation of a full load.

85

One example of this is an airport shuttle bus that transports passengers to rental car agencies. The car rental personnel are single-unit processors who can handle one customer at a time. If the shuttle delivers a dozen customers at once (a batch), most will waste a lot of time waiting in line. The same thing happens when a large quantity of parts emerge from a batch tool (like a heat treatment oven) simultaneously and arrive at a single-unit process (like a machining tool). This is, in fact, a transfer batching problem.

Hopp and Spearman discuss *cart sharing*, which allows workstations that produce identical products to receive inputs from and deliver their outputs to shared carts or hand trucks.[13] A similar technique might be considered by car rental agencies that are geographically close to one another: share their shuttle buses even though they are competitors. Thus, six shared buses that arrive and depart at ten-minute intervals will deliver, on average, half as many customers to each agency as would three buses that circulate at 20-minute intervals (Figure 4-3). The effect is to smooth the arrival of work for both agencies without requiring more buses. In addition, whatever time the customers lose by having the bus stop at two agencies instead of one should be more than made up by the shorter waiting times to catch a bus.

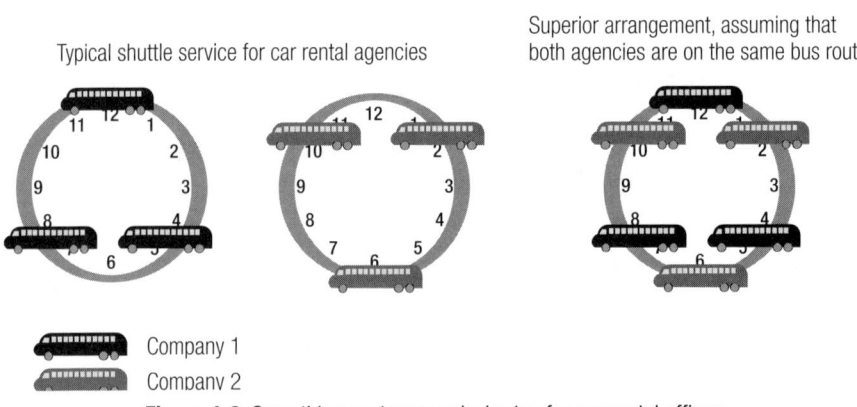

Figure 4-3. Smoothing customer arrival rates for car rental offices.

As Figure 4-2 illustrates, the first part emerges from the single-unit workstation at $t = 1$ but cannot enter the batch station until $t = 5$, so it waits 4 minutes. The second part must wait 3 minutes, and so on. The total waiting time is 4 + 3 + 2 + 1 = 10 minutes, or an average of 2 extra minutes for the five parts. The added cycle time is, in fact, $t\sum_{i=1}^{n-1} x = \frac{n(n-1)}{2} t$ when a batch process requires a load of n pieces that come from a single-unit process with a process time of t. The average extra cycle per piece is then $\frac{1}{n} \frac{n(n-1)}{2} t = \frac{(n-1)}{2} t$. If the work were moving in the other direction, something similar would happen when five parts arrived at a workstation designed to handle only one at a time.

Goldratt and Cox point out in *The Goal* that it is not necessary to wait for a full load unless a parallel process batch operation is a constraint, in which case anything less than a full load wastes irreplaceable capacity. It is nonetheless clear that batching usually adds non-value-adding cycle time to the process.

Henry Ford recognized that rush hour and all the traffic problems that come with it is the result of "batching," in which huge numbers of people arrive at or leave work simultaneously. The result was that he did not run his factory in shifts; arrival and departure times were staggered in 30-minute intervals so people were arriving and leaving throughout the day.

Parallel Process Batch Processes Are Harder to Control

Traditional statistical process control (SPC) assumes that the process is $N(\mu,\sigma^2)$, which is shorthand for "normal distribution with mean μ and variance σ^2." In this case, the control limits for the average of a sample of n pieces are $\mu \pm 3 \frac{\sigma}{\sqrt{n}}$. Control limits for the sample standard deviation or sample range, both of which reflect process variation, are also fairly straightforward. The process capability index, which measures the process's ability to meet the upper and lower specification limits, is $Cp = \frac{USL - LSL}{6\sigma}$, a ratio of the specification width to the process variation. Batching, however, introduces batch-to-batch variation that also must be accounted for as follows:

The mean of each batch follows the distribution $\mu_{batch} \sim N(\mu_{process}, \sigma^2_{between\_batch})$

Then for each individual measurement from a given batch of parts, $x \sim N(\mu_{batch}, \sigma^2_{within\_batch}) \Rightarrow x \sim N(\mu_{process}, \sigma^2_{within\_batch} + \sigma^2_{between\_batch})$ and therefore:

$$Cp = \frac{USL - LSL}{\sqrt{\sigma_{within\_batch}^2 + \sigma_{between\_batch}^2}}$$

where USL and LSL are the upper and lower specification limits, respectively.

Now for the average of n measurements from a batch,

$$\bar{x} \sim N\left(\mu_{process}, \frac{\sigma_{within\_batch}^2}{n} + \sigma_{between\_batch}^2\right)$$

And then control limits for the x-bar chart are,

$$\mu_{process} \pm 3\sqrt{\frac{\sigma_{within\_batch}^2}{n} + \sigma_{between\_batch}^2}$$

A computer can handle the more complex calculations, although the underlying behavior is likely to confuse shopfloor personnel. A problem that cannot be avoided is the huge uncertainty in the capability indices, which rely on accurate estimation of the variation. A process capability study is almost meaningless without a basis of 30 measurements, and 100 or more are preferable. Thirty or more *batches* (not pieces) are necessary to get even a mediocre estimate of the between-batch variation.

Automatic process control, which is often used in the chemical process industry, is also more difficult for batch processes. Continuous-flow chemical process equipment is easier to control than batch equipment, and the same applies to heat treatment furnaces. A classic Taguchi robust design problem involves a kiln for firing ceramic tiles.[14] The tiles on the inside of the load may finally reach the desired temperature but not as quickly as those on the outside, much as the inside of a large piece of meat will remain raw long after the outside is cooked. In the case of the tiles, this may or may not affect the final quality, but it is a very basic premise of manufacturing that all workpieces should experience identical process conditions. A Taguchi robust design was, in fact, necessary to fix quality problems that resulted from this arrangement. (This meant reformulating the tiles to make them less sensitive to the temperature variations inside the kiln.)

Alternatives to Batch Processing

A belt furnace, in which the parts travel through the heated zone on a conveyor belt, is a form of single-unit processing, and it introduces far less process variation. This approach was adopted at Ford's Highland Park plant more than 90 years ago.

> The blank is stiff from the finishing rolls of the plate-steel mill, and is placed at once on a steel-roller gravity-incline and carried to the annealing ovens, 1,500 degrees F [816°C], where the blanks are piled, one hundred and fifty on each of the six oven cars, which fill the oven with nine hundred blanks . . . the six cars go into the annealing oven at 8:00 P.M., remain until 12:00 midnight, are then withdrawn and left in piles on the cars until 6:30 A.M., where they are cool enough to be worked in the drawing press.[15]

In other words, the annealing oven accepted a process batch of 900 pieces, which took four hours to anneal. The pieces then had to cool for another six and a half hours before they could go to the drawing press. The Ford Motor Company was already quite familiar with the virtues of single-unit flow because:

> This first annealing practice will soon be obsolete. A furnace now under construction is served by an endless chain moving up and down, which is fitted with pendulum blank-carriers to take the blanks individually as they come through the press die, carry them upward about 60 feet [18.3 m] in the furnace uptake, giving ample heating time, and then carry the blanks downward 60 feet in the open air, giving plenty of cooling time before the blanks reach the oiling table.[16]

Modern technology provides additional options for converting batch processes to single-unit flow. Induction hardening uses an electric field to heat the workpiece, and it can process a part in a few seconds.[17] Lasers, high-frequency resistance hardening, flame hardening, and electron-beam hardening are additional possible single-unit alternatives to heat treatment.

Typical applications for such techniques include pieces that are only to be surface-hardened, and the rest of the workpiece quenches the hardened region by absorbing the heat. In high-frequency resistance hardening, high-frequency current heats a line of material at the part's surface to 1,600°F

(870°C) within half a second.[18] The steel that surrounds the heated stripe then quenches it immediately. The reference adds that cycle times for induction hardening, which can heat the entire piece or only its surface, are "a matter of seconds, and machines can be completely automated."

The single-unit heat treatment equipment can often be placed in line with the rest of the process tools (for example, in an assembly line or work cell), thus eliminating the need to move the work to and from a large oven somewhere else in the factory. The principle was well established at the Ford Highland Park plant more than 90 years ago, as shown by the placement of brazing furnaces right in the assembly line.

> Most of all, however, the Ford engineers have taxed the convolutions of their brain surfaces to shorten the lines of natural work-travel on the factory floors, first by crowding machine-tools together far closer than I have elsewhere seen machine-tools placed, and next by first finding the shortest possible lines of production travel of every car component, integral or assembled, and then placing every production agent needed either directly in that shortest line, or as near that line as possible, to the extent of placing even the brazing fires where most travel-saving advantage demands.[19]

Omark, a Canadian chainsaw manufacturer, adopted another alternative by selecting pretempered steel that did not require heat treatment at all. Heat treatment is, however, far from the only process that encourages or requires batching. Long setup times have always been an incentive to process large batches, and the following section shows how to deal with this problem.

SINGLE-MINUTE EXCHANGE OF DIE

The well-known formula for economic order quantity (EOQ) is $Q^* = \sqrt{\frac{2DS}{H}}$, where Q^* is the optimal order size, D is the demand, S is the setup cost, and H is the inventory holding cost. Heizer and Render show a similar Production Order Quantity Model,[20] but the optimal lot size is proportional to the square root of the setup cost in both cases. If the cost of setting up a tool to process a batch of parts is measured in time, it is clear that long setup times encourage the production of large batches. This is why manufacturing managers often hate the idea of "breaking the setup" to process a different product.

This is also why single-unit processing and mixed-model (A-B-C-A-B-C) production require very short setup times, which is achieved through single-minute exchange of die (SMED). The name is somewhat deceptive because it also applies to processes that do not use dies. The goal is to minimize non-value-adding setup time by changing the tool over more quickly and by externalizing parts of the setup job. *Internal setup* is setup work that requires the tool to stop, while *external setup* can be done while the tool is working on something else. Although SMED is most closely associated with Shigeo Shingo, American manufacturers were using it almost a century ago.

> In a certain shop with which we are familiar, a piece had to have several holes of different sizes drilled in it, a jig being provided to locate the holes. The drills and the sockets for them were given to the workman in a tote box. The time study of this job revealed several interesting facts. First, after the piece was drilled, the machine was stopped, and time was lost while the workman removed the piece from the jig and substituted a new one. This was remedied by providing a second jig, in which the piece was placed while another piece was being drilled in the first jig, the finished one being removed after the second jig had been placed in the machine and drilling started.[21]

This is a good place to introduce an important lean manufacturing concept: "There is far too much *muda* [waste, often wasted time] between the value-adding moments. We should seek to realize a series of processes in which we can concentrate on each value-adding process—Bang! Bang! Bang!—and eliminate intervening downtime."[22] A good way to envision this concept is to think of a stamping press that makes an actual "Bang!" when it strikes a part. This is the value-adding moment and everything else, such as moving the work to and from the press, is waste.

The Japanese are quite familiar with golf, which provides another way to explain the concept. Per Standard and Davis,[23] the golf club is actually in contact with the ball for 0.02 seconds. Everything that leads up to that value-adding "Bang!" is either transportation (walking or riding in the golf cart), waiting, or setup (swinging the club). In a four-hour, 90-stroke game, the golf club (tool) is in contact with the ball for a total of 1.8 seconds. Golf's purpose is recreation and exercise, and walking is a beneficial part of it. A factory, however, is in business to make money, so it is important to eliminate non-value-adding transportation and waiting times.

Hopp and Spearman point out that changing setups is very expensive in terms of lost time (which translates to marginal revenue) when the equipment is running at or near capacity but not when there is excess capacity.[24] In Theory of Constraints terminology, breaking a setup at a constraint is extremely expensive, but the cost of doing so elsewhere is essentially zero. This means that it costs little or nothing to smooth the production stream through mixed-model production at nonconstraints.

As noted above, American firms were using SMED at the beginning of the 20th century. Its origins go back much further, however, when the value-adding action was a literal bang: a shot from a black powder firearm.

Military Origins of SMED

Many lean manufacturing principles have military origins. This is not surprising because there is nothing like being shot at to encourage people to develop more efficient ways to shoot back. Soldiers created value only when they fired at the enemy; as with the stamping press, the value-adding "Bang!" can be taken literally. Everything else consisted of setup as preparation for firing a volley. Frank Gilbreth, the father of motion efficiency, wrote that his observations of military drills inspired him to apply the same principles to civilian industries.

> The U.S. government has already spent millions and used many of the best of minds on the subject of motion study as applied to war; the motions of the sword, gun, and bayonet drill are wonderfully perfect from the standpoint of the requirements of their use. This same study should be applied to the arts of peace.[25]

The loading drills for the muzzle-loading muskets and rifles that were in common use before the mid-nineteenth century may well have inspired Gilbreth. Most people have a mental image of a man, probably in a coonskin hat, measuring out powder from a powder horn and pouring it down the barrel. Hunters probably used this procedure because, if they fired and missed, the game would probably be out of range long before they could reload by any method. The procedure was far too slow, however, when the target shot back instead of running away.

Measuring out the powder charge is *internal setup* because the gunner cannot do anything else while he is doing it. Armies recognized this even in the era of the matchlock arquebus and musket (the 16th and 17th centuries), and soldiers were issued tubular wooden cartridges with premeasured charges. Premeasuring the charges externalized this aspect of the setup process and allowed the musketeer to fire more rapidly.

A matchlock gunner had to remove the burning match cord from the lock before he primed the pan unless he wanted to risk having the primer flask explode in his hand. Then he had to put the match cord back into the lock before he could fire. This is yet another example of internal setup because it required a shutdown of the tool in question. The flintlock eliminated this part of the setup activity by removing the live ignition source.

Paper-wrapped cartridges of the flintlock era also reduced setup time. The soldier began by biting open the cartridge, as this was faster than tearing it open with his hands. He poured some of the gunpowder into his musket's pan, and then dumped the rest down the barrel. The ball and the paper (the wad) followed, and a single push of the ramrod was enough to seat the bullet against the charge. Firing rates of four or even more rounds a minute were possible.

The ramrod itself, however, was another problem. Drawing it from its socket and then returning it added non-value-added setup time to the loading process. The movie *Sharpe's Rifles* depicts a highly improbable setup reduction technique: placing the bullet in the barrel and then slamming the weapon's butt on the ground, thus eliminating the need to draw and replace the ramrod. This *might* have worked for the smoothbore Tower Musket (Brown Bess), but it would have been hopeless to try to load the Baker Rifle's tight-fitting bullet in this manner. Moreover, it would have been potentially hazardous even for the Tower Musket, since failure to seat a muzzle-loading weapon's bullet against the powder charge can cause the barrel to burst upon firing. The fact that the technique would not have worked does not, however, take away from the lesson it teaches: Eliminate non-value-adding motions to gain a competitive advantage.

It was undoubtedly, however, every weapon designer's dream to combine the Tower Musket's ease of loading and high rate of fire with the rifle's accuracy. The problem was that the bullet had to go in with a loose fit and come out with a tight one, and the solution was provided by Captain Claude Miniè in the 1840s. It took but a light tap of the ramrod to seat the hollow-based Miniè bullet in a muzzle-loading rifle, but the pressure from the gunpowder charge expanded the base so the bullet would grip the rifling.

The contemporary von Dreyse rifle loaded from the breech, thus eliminating completely the need for a ramrod. The cartridge also contained its own primer cap, so the soldier no longer had to prime the weapon before he could fire. All these improvements in firearm design used principles that are recognizable as single-minute exchange of die, and Table 4-3 shows how

Table 4-3. Evolution of firearms and elimination of setup steps.

Matchlock Musket	Flintlock Musket	Von Dreyse Rifle
Remove burning match cord from lock	Pull the hammer back to half-cock (much faster than handling the match cord)	Open the breech bolt
Take out primer flask	Take a paper-wrapped cartridge from the pouch	Take a self-contained cartridge from the pouch
Open the flask and prime the pan	Tear or bite the cartridge open and pour some of the powder into the pan	Put the cartridge into the breech
Take out preloaded wooden cartridge	Pour the rest of the charge down the barrel and push the bullet and paper after it. (The paper forms the wad.)	
Pour the charge into the barrel		
Take a bullet from the belt pouch and put it into the barrel		
Draw the ramrod	Draw the ramrod	
Ram the charge	Ram the charge	
Return the ramrod	Return the ramrod	
Replace the match in the lock	Pull the hammer back to full cock (much faster than handling the match cord)	Close the breech bolt
The weapon is now ready to fire		

each successive step eliminated non-value-adding setup. The same thought process carries over into manufacturing and even service operations.

Additional SMED techniques include the use of split-thread bolts and quick-clamping flanges to connect tools quickly. The first of these seems to have originated with an innovative 19th-century design for artillery breech blocks, and its explicit purpose was "rapid setup."

Split-Thread Bolts and Quick-Clamping Flanges

Breech-loading artillery was a goal of weapon designers ever since the invention of gunpowder, and there are examples of crude breech loaders from medieval times. The need for a gas-tight seal at the breech was a practical obstacle until the invention of the interrupted-thread breech block in the 19th century. This meant that, instead of having to be screwed into place—a process that would probably have taken as long as loading from the muzzle—the breech block could be closed with a one-sixth turn.

> The French themselves had decided to change to breech-loading but they had adopted a hinged threaded block, without the separate breech-block that had caused all the problems in the Armstrong gun. The threaded block took so many turns to open and shut that the logical improvement was to cut away every sixth part of the threads; it needed only a one-sixth turn to shut it and yet it retained the strength of the fully threaded block.[26]

This example again reinforces the idea that the solution to a complicated problem is often obvious *only in retrospect*. This is why no job or procedure should ever be taken for granted, with the assumption that no improvement is possible.

The interrupted thread or split-thread bolt (Figure 4-4), which is useful for clamping industrial equipment with a one-quarter or one-sixth turn of the wrench, uses exactly the same principle. The difference between securing a die or other piece of tooling with bolts that must be turned a dozen or more times (even with a power tool) and bolts that can be tightened with a quarter or sixth turn of the wrench is enormous, and it is this kind of thinking that yields enormous reductions in setup time.

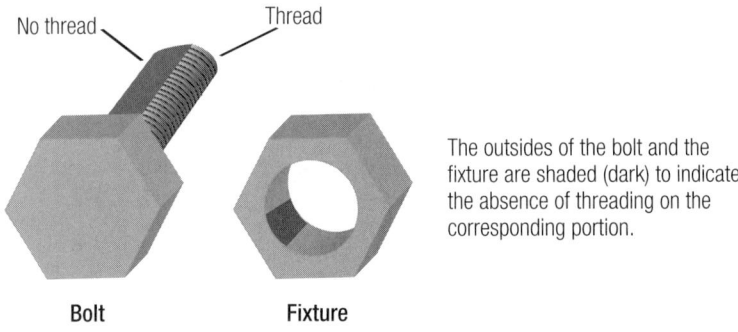

Figure 4-4. Interrupted thread or split-thread bolt.

Another way to reduce the turns needed to secure a bolt involves a clamp with a pear-shaped hole. The bolt is always left almost all the way in the female threads. To secure a new piece of equipment, the large part of the "pear" fits over the head of the bolt. A short sliding motion moves the narrow part of the "pear" under the bolt's head, and a single turn of the wrench is usually enough to tighten the assembly (Figure 4-5).

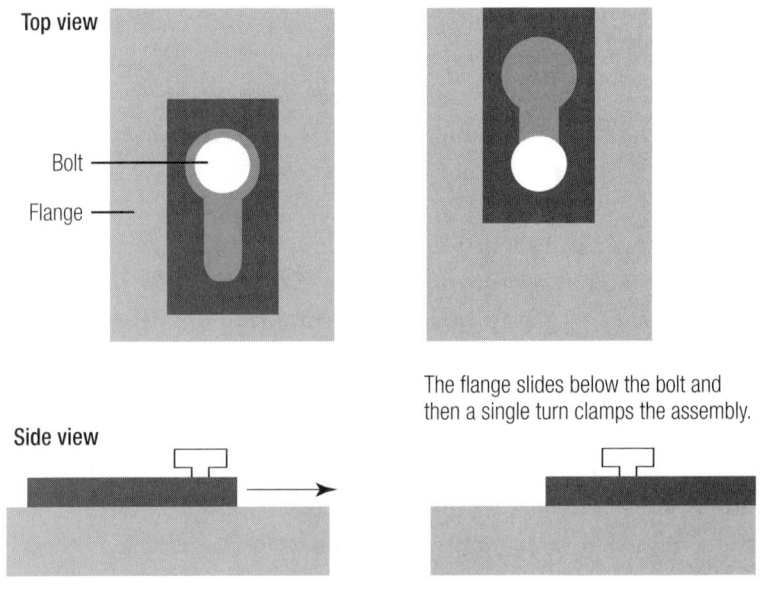

Figure 4-5. Quick-clamping flange.

In summary, the key concepts behind SMED are (1) changing internal setup to external setup, and (2) minimizing all non-value-adding motion that is associated with clamping and securing tooling. This section has covered the virtues of single-unit processing and techniques to reduce the need for batching. The following section covers work cells and unitary machines, which also help reduce processing time and variation in processing time.

WORK CELLS AND UNITARY MACHINES

This book has already described how the Ford Motor Company spaced process equipment as tightly as possible more than ninety years ago. The idea was for the work to move directly from one workstation to the next without going offline to a storage area or even to a hand truck for intraplant transportation. Transportation and storage add cycle time (waste) and also time variation that prevents 100 percent utilization. The spaghetti diagram is a recognized analytical tool for assessing the distance that the work must travel.

Spaghetti Diagram

Readers of *The Family Circus* comic strip, in which a boy often runs all over the neighborhood to get to the house next door, will appreciate the value of the *spaghetti diagram*. It gets its name from the drawing of lines to track a process routing through the factory. Like the boy in the comic strip, parts may travel thousands of feet (or even a few miles) before they get to the shipping dock.

Heizer and Render quantify the costs that are associated with the spaghetti diagram as follows. Cost = $\sum_{i=1}^{n}\sum_{j=1}^{n} X_{ij} C_{ij}$ C_{ij} is the cost to move a lot or batch between departments i and j, and X_{ij} represent the number of lots to be moved.[27] It is important, however, to consider the cycle time costs in addition to the monetary costs of moving the pieces. The unquestioning use of the preceding metric may, in fact, encourage dysfunctional behavior like letting the parts wait until they make up a full forklift-load or cartload. For example, it "costs" twice as much on paper to move two ten-piece lots as it does to move a twenty-piece lot, and acting on this observation encourages batching and queuing as opposed to single-unit flow. In general, it might be better to focus on reducing the cost (whether in time, money, or both) of moving lots to encourage lot splitting or even single-unit flow. The moving assembly line and the work cell realize this goal by reducing the movement cost to virtually nothing.

Work Cells

The work cell and, even more so, the unitary machine dramatically reduce material transfer time and the accompanying variation. U.S. entrepreneurs invented cellular manufacturing almost 100 years ago, although it was not known by that name at the time. Ford's *My Life and Work* describes the principles of cellular manufacturing very explicitly along with its application at the Highland Park factory. Ford's principles of assembly are as follows:

1. Place the tools and the men in the sequence of the operation so that each component part shall travel the least possible distance while in the process of finishing.
2. Use work slides or some other form of carrier so that when a workman completes his operation, he drops the part always in the same place—which place must always be the most convenient place to his hand—and if possible have gravity carry the part to the next workman for his operation.
3. Use sliding assembly lines by which the parts to be assembled are delivered at convenient distances.[28]

Although the Ford Motor Company was the unquestioned birthplace of what we now call lean manufacturing, it probably adopted principles that others had recognized even earlier:

Only by relating each machine with the others in such a way that production will follow in straight lines without confusion, can the highest economy of operation be attained … In this machine shop of the Mueller Machine Tool Company, [floor space] is well utilized by arranging the machines logically with respect to production.[29]

H. L. Gantt, the inventor of the Gantt chart, adds, "In another shop machinery was rearranged so as to bring together allied operations and reduce the time of transportation."[30]

Why should production "follow in straight lines without confusion?" The U-shaped cell, with the work traveling counterclockwise, often results in better ergonomics. This is especially true if a worker handles more than one machine and must walk between them. In 1911, however, the tooling layout had to be in straight lines because the overhead rotating shafts from which the tools drew their power were straight (Figure 4-6).[31] It was, in fact, Henry Ford's widespread introduction of an individual electric motor for each tool that allowed the invention of the U-shaped cell. Nonetheless, U.S. manufacturers had clearly recognized the basic principle of arranging the tools in the production sequence no later than 1911.

Figure 4-6. Early 20th century factory; tools draw power from overhead shafts.

With this innovation, a typical work cell contained all the equipment necessary to make any part in a given production family:

> As the factory is now organized, each department makes only a single part or assembles a part. A department is a little factory in itself. The part comes into it as raw material or as a casting, goes through the sequence of machines and heat treatments, or whatever may be required, and leaves that department finished.[32]

Thus, each work cell had the following key characteristics:

- **All tools necessary to make the part are placed in the sequence of operations.** Ford placed brazing operations directly in line with other

operations. This eliminated the need to transport the work to and from a separate brazing department. Heat-treatment, a frequent cause of batching and queuing, can also be done in line. Possible methods include high-frequency resistance hardening, flame hardening, induction heating, laser hardening, and electron beam hardening.
- **Tools are close enough to allow the work to move by slide or conveyor, as opposed to requiring transportation by a forklift or hand truck.** As with the fish-cleaning example from *Captains Courageous*, this approach eliminates the non-value-adding activity (transportation) along with all time variation that goes with it.

Ford's work cells made only one part (although they were designed so they could make others simply by changing the dies). *Group technology* means designing work cells to make families of parts. All members of a *production family* can be produced by the same sequence of operations, which means that the same work cell can produce whichever part happens to be needed. Group technology is synergistic with single-minute exchange of die because SMED allows a work cell to switch from one production family member to another without evoking complaints about "breaking the setup."

If a work cell places all the necessary production tools side by side, the unitary machine, discussed in the following section, eliminates even the work slide, conveyor, or rollway by performing all the necessary operations on one machine.

The Unitary Machine

A *unitary machine* essentially combines all of a work cell's operations into one machine. Ford cited the turret lathe, which could bore, ream, mill, and drill, as an example. "Thus there is no lost time or motion; the parts are handled but twice—once in loading as a rough forging and again in removing as a finished part."[33] Note the observation that the parts are handled only twice. Loading the work into the tool and possibly clamping it into place adds no value while creating an opportunity for handling damage. The System Company goes so far as to say that handling an item twice is like buying the same material twice.[34] *Beyond the Theory of Constraints* recommends that, for the purposes of value stream analysis, "processing" should be broken down into handling and actual value-adding work like machining.

Otherwise, the waste of handling can hide in plain sight as "processing," and nothing will ever be done about it.

We have already seen how military necessity drives improvement, and the first unitary machine may have been a weapon:

> The [Gatling gun's] cartridges were gradually fed into the chamber by cams as the barrels revolved, then fired at the bottom position, and extracted and ejected during the upward movement. As the barrel reached the top, it was empty and ready to take the next round. The great advantage of this system was that it divided up the mechanical work between the six barrels so that the machinery was operating at a reasonable speed, and it also allowed each barrel time to cool down between shots.
>
> ... the plunger for a particular barrel would start at the fully drawn position. A cartridge was dropped into the groove ahead of the plunger and, as the unit revolved, the plunger moved forwards and rammed the cartridge into its barrel, fired it, and then extracted the case and ejected it through the casing.[35]

These statements clearly show the thought process behind the unitary machine. The Gatling gun, in fact, went beyond the unitary machine by breaking the "operations" down into differential elements. That is, while a unitary machine must pause while discrete steps like boring, reaming, milling, and drilling take place at each station, the Gatling gun's operations of loading, ramming, the value-adding literal "Bang!" and extraction took place while the machine and the "parts" were in continuous motion.

The virtue of this approach is evident when one considers that even a hand-cranked Gatling gun could fire 600 rounds per minute, and Gatling's addition of an electric motor in the late 19th century raised this to 3,000 rounds per minute. It should also be noted that Gatling guns jammed rarely, if ever; these "unitary machines" were not prone to stoppages. Although applying the Gatling gun's principles to the design of production equipment may be somewhat more difficult, the approach might be worth keeping in mind.

Recall the plug flow reactor, which was described in Chapter 2 as an ideal production model. It is somewhat analogous to the Gatling gun because each unit—in this case, a cylindrical fluid element of differential length—is processed continuously as soon as it enters the equipment. The fluid element contains all the necessary reactants, in the right proportions. It is as if the fluid element is a kit that contains the entire bill of materials. The reaction (processing) continues without pause as the fluid element moves through the pipe.

The first civilian unitary machines probably appeared in the early 20th century. Arnold and Faurote describe one that they observed at Ford's Highland Park plant:

> The top of the four-arm indexing fixture turns toward the operator. . . . The top of the fixture is the put-on-and-take-off station. As the fixture is indexed round, the piston first comes to the drilling station, and has the pin-hole drilled. At the second station, a second piston is drilled while the first pin hole is being bored. At the third station, a third piston is being drilled while the second piston is being bored and the first piston is being reamed. The next movement of the fixture brings the first piston to the top position, where the attendant removes it, and replaces it with another, and from thence on, so long as the attendant removes and replaces the pistons, all three operations of drilling, boring, and reaming piston pin-holes are in continuous progress, save for the time occupied by drawing the tools back and indexing the fixture from one position to the next following position.[36]

The work cell, unitary machine, and moving assembly line (in which all activities are also in the sequence of operations) differ enormously from the departmental or "farm" layout, in which all machines of a certain type are located in one part of the factory. The departmental layout encourages batching (and all the problems that go with it) for convenience in moving parts from one department to another. Transportation of parts, whether by forklift or hand truck, is waste. It requires labor and adds cycle time to the process so it should be avoided if possible. Arnold and Faurote recognized this quite explicitly more than 90 years ago:

> Of course, after what has been here said, the visitor will not expect to find in the Ford shops any examples of orthodox machine-tool placing in generic groups, lathes together in one place, drilling

machines, milling machines, and planing machines each in a group by themselves.[37]

Dedicated Equipment by Product Family

The cellular manufacturing concept is useful even in shops where equipment cannot actually be organized into cells. This may be the case, for example, when immovable pieces of equipment (called *monuments*) are present, or when other considerations prevent physical movement of the tools.

In this case, specific tools can be dedicated to specific products (or product families). Then the products can be routed through their dedicated tools so that the tools will *act* like a work cell even if they are not in physical proximity to one another.[38] The only problem is, of course, the need to move the work from one area to another, which may necessitate accepting some transfer batching and work-in-process.

SUMMARY: VARIATION REDUCTION

This chapter treats numerous ways to suppress variation in processing and material transfer times. Subdivision of labor, as practiced by Henry Ford, eliminates not only non-value-adding handling activities but also the variation that goes with them. The purpose was not to make the work mindlessly repetitive, but to relieve the worker of boring actions for which he could not even be paid.

Value stream analysis is a well-established technique for classifying activities as value-adding, value-assisting, or non-value-adding. The Automotive Industry Action Group (AIAG) recommends six categories, while Frank Gilbreth identifies four. This book suggests five as the minimum necessary to force all non-value-adding work to become visible: (1) transform, (2) handle, (3) transport, (4) inspect, and (5) delay. It is necessary to identify handling activities like picking work up and putting it down, whether by a person or a machine, to prevent these non-value-adding actions from masquerading as "processing."

This book uses *time accounting* in the sense of measuring cycle time (as opposed to labor and professional services). The purpose of time accounting is to force wasted cycle time to become visible. The idea of designing processes that make workers stand idle 95 percent of the time would appall any businessperson, but processes frequently allow parts to do just that.

Batching, whether for processing or transfer, aggravates time variation. Process batching also adds to process variation and makes statistical process control more difficult. Batching has numerous vices; indeed, its only virtue is convenience in material movement. The moving assembly line and its analogues address the latter issue so factories should strive for single-unit, or at least small-lot flow, whenever possible. Reluctance to break setups is a major cause of serial process batching. Single-minute exchange of die addresses this issue and facilitates mixed-model production.

Henry Ford's moving assembly line placed all the necessary tooling for a given job in series and in close proximity. This approach allowed single-unit flow with little loss of time in moving work from one tool to the next. Placement of the necessary tooling for a given routing in line, instead of in separate departments, ties in with group technology, as it should be possible to use the same series of tools to make any part in a given production family.

Ford's replacement of overhead power shafts with individual electric motors for each tool made the modern U-shaped work cell possible. The unitary machine, which needs to pick up and release the part only once while performing multiple operations, is an extension of the work cell concept.

If there are practical barriers to the organization of tools into work cells, specific tools can be dedicated to specific production families. The tools then emulate a work cell, although material movement remains a problem.

Chapter 5 discusses methods for improving productivity (which includes elevating the constraint) and reducing cycle time.

Endnotes

1. Shakespeare, *Julius Caesar*, Act I, Scene 2.
2. Ford, *My Life and Work*, p. 83.
3. Ibid., p. 82.
4. Hradesky, *Productivity & Quality Improvement*, p. 27.
5. Levinson, *Henry Ford's Lean Vision*, p. 255.
6. AIAG, *CQI-10: Effective Problem Solving*, p. 193.
7. Shingo, *Zero Quality Control*, p. 162.
8. Robinson, ed., *Modern Approaches to Manufacturing Improvement*, pp. 85–86.
9. Ohno, *Toyota Production System*, p. 17.
10. AIAG, *CQI-10: Effective Problem Solving, A Guideline*, p. 79.

11. Hopp and Spearman, *Factory Physic*, pp. 305–306.
12. Shingo, *Sayings of Shingo*, p. 162.
13. Hopp and Spearman, *Factory Physics*, p. 598.
14. Quality Council of Indiana, *Certified Six Sigma Black Belt Primer*, XI-14 to XI-15.
15. Arnold and Faroute, "Ford Methods and the Ford Shops," pp. 86–87.
16. Ibid., p. 0.
17. Cubberly and Bakerjian, *Tool and Manufacturing Engineers Handbook*, pp. 41-18 to 41-21.
18. Ibid., p. 0.
19. Arnold and Faroute, "Ford Methods and the Ford Shops," pp. 38–39.
20. Heizer and Render, *Production and Operations Management*, pp. 582–583.
21. Robert Thurston Kent, introduction to *Motion Study*, by Frank Gilbreth.
22. Imai, *Gemba Kaizen*, pp. 22–23.
23. Standard and Davis, *Running Today's Factory*, p. 61.
24. Hopp and Spearman, *Factory Physics*, p. 63.
25. Gilbreth, *Motion Study*.
26. Preston, *Battleships*, p. 24.
27. Heizer and Render, *Production and Operations Management*, pp. 394–395.
28. Ford, *My Life and Work*, p. 80.
29. The System Company, *How to Get More Out of Your Factory*, pp. 124–125.
30. The System Company, *How Scientific Management is Applied*, p. 21.
31. Ibid, p. 55.
32. Ford, *My Life and Work*, pp. 83–84.
33. Ford, *Moving Forward*, pp. 130–131.
34. *How to Get More Out of Your Factory*, p. 50.
35. Hogg, Ian. *Guns and How They Work*, p. 63.
36. Arnold and Faroute, "Ford Methods and the Ford Shops," pp. 208–209.
37. Ibid., pp. 38–39.
38. See p. 441 of *Factory Physics* by Hopp and Spearman for more on this concept.

FIVE

Productivity Improvement

Chapter 4 ended with a discussion of how work cells, moving assembly lines, and unitary machines all place the necessary tooling in the correct sequence for a given process. This approach eliminates the waste and variation involved in moving work from one department (for example, stamping) to another (like grinding, milling, or heat treatment). This chapter overlaps somewhat with Chapter 4 because many productivity improvement techniques also reduce variation in processing and material transfer times. The current chapter first addresses motion efficiency, continues with in-line quality control techniques such as self-check systems and error-proofing, and finally examines the role of the supply chain.

The concept of friction, waste, or muda is a prerequisite for a true understanding of lean manufacturing and productivity improvement. An enormous amount of waste can hide in plain sight because people become used to living with or working around it.

FRICTION

Friction is a term from Carl von Clausewitz's *On War*, and it sums up the impact of waste, or muda, on military activities.

> Everything in war is very simple, but the simplest thing is difficult. The difficulties accumulate and end by producing a kind of friction that is inconceivable unless one has experienced war. . . . Countless minor incidents—the kind you can never really foresee—combine to lower the general level of performance, so that one always falls short of the intended goal.

... Fog can prevent the enemy from being seen in time, a gun from firing when it should, a report from reaching the commanding officer. Rain can prevent a battalion from arriving, make another late by keeping it not three but eight hours on the march, ruin a cavalry charge by bogging the horses down in the mud, etc.[1]

The same principles carry over into industry, where a key characteristic of friction is its seemingly minor nature. Major scrap and rework incidents, missed deliveries, and catastrophic equipment breakdowns usually get attention, and their underlying causes are corrected. Friction usually does not get that sort of attention because people become used to living with it or working around it. J. F. Halpin defines friction as "the little things that get under a worker's skin but are never quite important enough to make him come to management for a change."[2] Tom Peters cites "the accumulation of little items, each too trivial to trouble the boss with" as a primary cause of miss-the-market delays.[3] Schonberger adds:

> The machine that jams sometimes, the tool that must be searched for sometimes, the assembler who does the task the wrong way sometimes, the part that arrives late sometimes, the blueprint that is wrong sometimes, the part that is off the mark sometimes—all of these and many more require costly sets of "solutions." They are not true solutions, because they provide ways to live with the problems.[4]

Henry Ford wrote essentially the same thing: "It is the little things that are hard to see—the awkward little methods of doing things that have grown up and which no one notices. And since manufacturing is solely a matter of detail, these little things develop, when added together, into very big things."[5] One of Henry Ford's principal success secrets was his ability to recognize friction and waste on sight. Harry Bennett describes how Ford recognized material waste that most observers would have overlooked:

> One day when Mr. Ford and I were together he spotted some rust in the slag that ballasted the right of way of the D. T. & I [railroad]. This slag had been dumped there from our own furnaces.

PRODUCTIVITY IMPROVEMENT

> "You know," Mr. Ford said to me, "there's iron in that slag. You make the crane crews who put it out there sort it over, and take it back to the plant."[6]

Ford's sensitivity to waste may have come from the fact that he grew up on a farm, where he had ample opportunity to observe wasteful farming practices firsthand:

> I believe that the average farmer puts to a really useful purpose only about 5 percent of the energy that he spends. If any one ever equipped a factory in the style, say, the average farm is fitted out, the place would be cluttered with men. The worst factory in Europe is hardly as bad as the average farm barn. Power is utilized to the least possible degree. Not only is everything done by hand, but seldom is a thought given to logical arrangement. A farmer doing his chores will walk up and down a rickety ladder a dozen times. He will carry water for years instead of putting in a few lengths of pipe. His whole idea, when there is extra work to do, is to hire extra men. He thinks of putting money into improvements as an expense. Farm products at their lowest prices are dearer than they ought to be. Farm profits at their highest are lower than they ought to be. It is waste motion—waste effort—that makes farm prices high and profits low.[7]

In other words, the farmer learns to live with wasteful practices like carrying buckets of water instead of installing a pipe to carry it because the job does get done and there is no immediate pressure for change. Proactive managers, on the other hand, must recognize the necessity or desirability of making improvements before circumstances force them to do so. This means they must also understand the concept of false economy and how it contributes to friction.

FALSE ECONOMY
False economy means saving pennies now only to lose dollars later, and it is the cause of endless trouble in many businesses. Toyota has already defined seven classes of muda or waste, and this section will similarly classify several

forms of false economy. It is useful to break these down into typical cause-and-effect diagram categories as follows:

- False economy of cheap equipment (machines); this includes buying cheap but non-capable instruments (measurements)
- False economy of cheap workers (manpower)
- False economy of cheap purchases (materials)
- False economy of cheap working conditions (environment)
- False economy of cheap procedures (methods)

False Economy of Cheap Equipment

An old adage states that a poor person must buy good quality. A rich person can afford to replace thin-soled shoes that wear out in a year or less, but a poor one cannot. Success in business comes from thinking from the poor person's perspective: specifically, that no business can afford waste. For example:

> Economy in the use of leather belting can only be attained by purchasing the best grades of belting, made by firms of established reputation, and then by applying it and caring for it in an intelligent and proper manner. It is an easy matter to buy belting for 10 or 12 percent less than is usually paid for first class goods. But it will be made from the leather cut too far from the center of the hide and consequently will have thin and soft spots, which, coming on the edge of the belt, will permit it to stretch unequally. If a piece forty feet long is laid on the floor, it will be impossible to make it conform to a straight line. Therefore, it will not run true on the pulleys, and if it is used on cone pulleys, the edges will turn up and the belt will soon be useless. In situations when a good belt would run a year and still be in good condition, this kind of belt will not last three months. It is the poorest of economy to save ten percent by putting in belts below the standard in quality.[8]

Another example of false economy was to run machine tools slowly to maximize the life of the tool. Frederick Winslow Taylor and Maunsel White, a metallurgist at Bethlehem Steel, developed steel alloys that allowed much higher speeds for cutting tools. Taylor pointed out that the principal goal was to cut metal, not to avoid wear on the tooling. Thus, the performance measurement should be metal removed per unit time, and tools should be reground as needed to achieve this.[9]

The return on investment metric also can promote the false economy of cheap equipment. Under traditional cost accounting, return on investment is profit divided by investment. This approach can encourage an organization to keep obsolete production equipment to avoid increasing the metric's denominator by investing in new equipment. Ford recognized clearly that "saving money" by not investing in new equipment is often a false economy:

> If a device would save in time just 10 percent, or increase results 10 percent, then its absence is always a 10 percent tax. If the time of a person is worth fifty cents an hour, a 10 percent saving is worth five cents an hour. If the owner of a skyscraper could increase his income 10 percent, he would willingly pay half the increase just to know how. The reason why he owns a skyscraper is that science has proved that certain materials, used in a given way, can save space and increase rental incomes. A building thirty stories high needs no more ground space than one five stories high. Getting along with the old-style architecture costs the five-story man the income of twenty-five floors. Save ten steps a day for each of twelve thousand employees and you will have saved fifty miles of wasted motion and misspent energy.[10]

Those are the principles on which the production of my plant was built up.

False Economy of Cheap Labor

During the first part of the twentieth century, when the development and spread of scientific management and lean manufacturing methods were at their height, every intelligent business leader knew that low-wage workers were expensive. Henry Ford stated explicitly:

> Good workmanship has to be paid for, and good workmanship is cheap at almost any price. It is simply a waste of time and money to

erect an elaborate manufacturing equipment and then expect that it can be run by low-paid men.

... Hiring men because they are cheap will ruin a business as quickly as buying material because it is cheap.[11]

It was Ford's policy to pay whatever wages were necessary to get the best sailors and officers for his cargo ships. His Great Lakes ore carriers had almost the kind of accommodations that first-class travelers might expect on a passenger ship. As Ford explained:

On the whole, our wages will run considerably higher than the highest wages elsewhere paid. We make money on these wages, for really, the whole total of wages paid on a ship is not very important—the important thing is to see that you get the full use out of the big investment, which is the ship. . . . Low-priced, irresponsible men will not care what happens to a ship or how long it stays in port.[12]

Basset adds:

We all know that cheap labor is not cheap. . . . In any operation in which the material costs are high as compared with the labor costs, the highest possible pay is the cheapest if it results in savings of material, or in a fine product, or in both. In the grades of production where labor is the big factor, high wages are economical if the wastes of human power can be cut to a minimum.[13]

It is interesting to speculate on the influence of the desire for cheap labor on the world's armies, which have often led civilian industries in improving efficiencies. The crossbow, and then the musket, provided military commanders with the unprecedented opportunity to hire cheap projectile troops. The longbow was a high-skill weapon that took years of training to master. In contrast, a sergeant could walk into the lowest-class tavern in any city and get what the Duke of Wellington called the scum of the earth to drink his monarch's health for a shilling (or its equivalent) a day. The aforementioned scum of the earth could then (again to use the Iron Duke's terminology) be made into fine fellows who could load muskets and, on command, point them in the enemy's general direction and pull the triggers. The arrangement was eminently sensible to every major European power during the pike-and-shot and horse-and-musket eras of warfare.

In retrospect, however, it can be argued that reliance on muskets was a false economy. It is especially telling that Benjamin Franklin recommended in 1776 that the Continental Army adopt the longbow.[14] An examination of this weapon's inherent capabilities shows that Franklin's advice to revive a weapon that had been out of style for more than two centuries was far more practical than quixotic.

As observed by Prince Louis Napoleon, "A first rate English archer who, in a single minute, was unable to draw and discharge his bow 12 times with a range of 240 yards and who in these 12 shots once missed his man, was very lightly esteemed."[15] In the hands of a skilled archer, the Welsh longbow would have certainly been dangerous to massed targets like shoulder-to-shoulder infantry formations at substantially greater distances. Its sole vice was its need for an archer who had spent his entire life learning to use it—that is, what Taylor and Ford would have called a "first-class man." In contrast, the sole virtue of the medieval crossbow and its successor, the matchlock arquebus, was the fact that almost anyone could learn to use them fairly quickly.

An enemy soldier had to be very unlucky to be hit by an arquebus at more than 100 paces, at least by one that had actually been aimed at him. The arquebus' smoothbore successors, namely the matchlock and flintlock musket, suffered from the same deficiency. All these weapons were best employed against massed targets, and the basic tactic was to fire as rapidly as possible in the enemy's general direction. Given Poisson's model for random arrivals, someone's luck had to run out sooner or later.

Firing as rapidly as possible was another issue entirely. The arquebusier (the person using the arquebus) had to start the loading process by removing the burning match cord from his weapon's lock to avoid blowing his hand off when he brought his priming flask to the pan. He then had to ram the charge and ball down the muzzle, replace the match cord, place the weapon on its forked rest, and await the command to fire. Upon doing so, he and the other men in his rank marched to the back (a process known as countermarching) and carried out the tedious reloading process while they made their way back to the front. Henry Ford pointed out that pedestrianism is not a highly paying line of work, and soldiers whose principal activity is to countermarch and reload inefficient weapons, as opposed to delivering a literal value-adding "bang," cannot earn high wages either.

It is therefore easy to envision a small body of longbowmen—perhaps a hundred against a thousand arquebusiers—conducting a one-sided massacre by simply standing out of arquebus range and delivering twelve shots per man per minute, with few if any arrows missing their targets. This, of course, presumes no interference by cavalry, but both sides would have had pikemen to keep enemy riders away.

Longbowmen would have been far more expensive in terms of pay, for one does not hire people like that for a shilling a day. Pay was, however, but a small part of the financial equation. A commander might easily hire ten thousand cheap men for a shilling a day, but then he had to feed them. This was a potential problem in regions where provisions might be unavailable at any price, and an ongoing expense under any circumstances. He might have done better to hire five hundred "first-class men" for a pound a day while saving considerable money on their upkeep. This also would have reduced his risk of having to abandon a campaign because of logistics problems.

By the time Franklin recommended adoption of the longbow, the flintlock musket had superseded matchlock weapons. The British Tower or Brown Bess musket, by virtue of its loose-fitting bullet, could fire upwards of four rounds per minute—but the same loose fit rendered the weapon useless at more than 150 paces. The longbow had twice the rate of fire and was effective at twice the range.

Today's manufacturers might find it advantageous to keep these military examples in mind when deciding between cheap labor and well-paid labor. They should consider carefully whether cheap labor would save money or waste it. In other words, they should understand the difference between a bargain and false economy. For example, manufacturers that hire cheap workers in China discover that their lead times become much longer if the product must travel by container ship and not air freight. This leads to the next topic, the false economy of cheap purchases.

False Economy of Cheap Purchases

This subject was initially introduced under the related subject of dysfunctional performance measurements in purchasing. When purchasing managers buy unnecessarily large quantities to get discounts, the result is often wasteful stores of inventory that might become obsolete.

Looking offshore for cheap parts also can be a false economy. A container ship is essentially a cleverly disguised floating warehouse that (1) invites batching and (2) adds enormous lead times to the supply chain. Furthermore, any quality problems in the shipment will not become evident for weeks.

Needless to say, materials and parts that are cheap because they are substandard will often create in-line or post-sale quality problems, and thus end up causing far more expenses than savings.

False Economy of Cheap Working Conditions

Readers of Charles Dickens' *A Christmas Carol* will recall how Ebenezer Scrooge skimped on coal even in the depths of winter to save money. This not only made his employee Bob Cratchit unhappy and uncomfortable, it probably also reduced both men's efficiency. Much of their work involved writing, and cold hands do not write well. Moreover, the ink probably did not flow smoothly either. Even if a product does not require tight temperature and humidity controls, it is worth spending money to provide comfortable and ergonomic working conditions for those who make it.

The wisdom of lighting factories and offices with cheap and energy-efficient fluorescent light can and should be questioned. Fluorescent light is not the white light for which the human eye is designed. The advantages of using the solar spectrum, or the closest equivalent, were recognized almost a hundred years ago.

> Good light to work by is an investment too infrequently made in the factory. In comparison with the cost of labor, the cost of artificial light is trifling, but there are thousands of skilled mechanics who lose efficiency because of insufficient light.[16]

The same reference mentions the problems that might arise from non-white light. A mercury vapor lamp was said to have "the peculiar and very desirable quality of diffusing light around corners, as it were." Office workers discovered, however,

> ... the disadvantage of all colors being changed, a red looking purple, a brown looking green, a blue looking pink, and so on ... Office help are apt to complain on account of its color effect, but shop workmen, particularly where the light is at a considerable height over their heads, express unqualified praise for the light and its soft penetrating qualities.[17]

Another hazard that was unknown at the time involved ultraviolet radiation from these lamps, which can harm the eyes and skin through long exposure. Moreover, such lighting is obviously less suitable for office work than shop work. Gilbreth adds:

> Light in a factory is the cheapest thing there is. It is wholly a question of fatigue of the worker. The best lighting conditions will reduce the percentage of time required for rest for overcoming fatigue. The difference between the cost of the best lighting and the poorest is nothing compared with the saving in money due to decreased time for rest period due to less fatigued eyes.[18]

False Economy of Cheap Procedures

Many organizations institute procedures that save money on paper while squandering money, cycle time, or even more important assets in practice. During the Second World War, it was common practice for bomber pilots to maneuver to avoid enemy antiaircraft fire.

> The groups, which had begun bombing before [Curtis] LeMay arrived, had been jinking (maneuvering sharply at random) to avoid flak as they approached their targets. This jinking maneuver continued through the bomb run. The result: The bombing ballistic solution was thrown off and the bombs were going everywhere but on the target.
> ... [Curtis LeMay] concluded that the accepted jinking maneuver gave an airplane no greater probability of avoiding flak than flying straight and level. When mission orders came to bomb St. Nazaire, France, on November 23, 1942, LeMay ordered his crews to fly a straight path for seven minutes to align sights and release

bombs. At this prospect, many of his crews, including some squadron commanders, thought he was crazy. But the mission proceeded, and LeMay flew in the lead plane.

The results were impressive: No losses due to flak. Two aircraft lost to fighter action. This was much better than the other groups flying to St. Nazaire that day.[19]

If the bombs hit their targets the first time, there was no need to risk a second mission. Even if flying in a straight line increased the risk of being hit, the reduction in necessary missions increased the aircrafts' and crews' survival chances while saving fuel and bombs as well.

Another example of false economy is the frequent proposal that the speed limit should be reduced to save gasoline. Some have even argued that trucks should drive more slowly to save diesel fuel.

> The Department of Energy estimates that every five miles per hour a person drives above 60 mph costs an extra 20 cents a gallon, for a fuel-efficiency loss of 7% to 23%, depending on the type of car and gas. That's because higher speeds increase aerodynamic drag on a car, requiring more horsepower. Over a year, it costs roughly an additional $180 in gas to drive 75 mph instead of 60 mph, according to the Environmental and Energy Study Institute, which promotes energy efficiency and renewable energy.[20]

This is an excellent example of false economy. Assume a 300-mile trip on an interstate highway in a full-sized car that gets 25 miles to the gallon. The trip requires 12 gallons of gasoline and five hours at 60 miles an hour. If it is legal and safe to drive 75 mph, the trip requires only four hours but incurs a penalty of 60 cents per gallon for a total of $7.20. By driving only 60 miles an hour to save gasoline, the driver finds himself working for $7.20 an hour in after-tax money.

Another "cheap procedure" involves making people wait at tool rooms or stockrooms to get parts and materials. Henry Ford discovered that it often took 25 cents' worth of a worker's time to get a 30-cent tool. To put this waste in perspective, quarters were made of silver in those days, and a dime could buy a loaf of bread.[21]

MOTION EFFICIENCY

The importance of motion efficiency in jobs with any significant labor component cannot be overemphasized. Assessment of the motions the worker must make to perform the task, through the use of a process flowchart, will often show many non-value-adding activities. It is often useful to videotape a job as it is being done, with the clear explanation that the purpose is to assess the job design and not the worker. Workers can even do this themselves, and they will often be amazed to see how much waste is built into a given task.

The joke about installing a light bulb by having one worker hold it while three others turn the ladder often seems far less funny after a production team finds analogous waste in the factory. The light bulb joke involves four people doing the work of one, with far more physical effort on the part of everyone involved. Edward Mott Woolley provides a wonderful, real-world example from a fabric folding operation in a bleaching and dyeing factory in Wilmington, Delaware: "But all [employees] took two steps to the right to secure their cloth, returned to the tables, folded the stuff and deposited it on another pile two steps to the left. That had always been the practice; no one had ever thought to question it."[22]

Although not as inefficient as the light bulb example, this job required two people to do the work of one simply because no one noticed the waste that was hiding in plain sight. The solution, which was to move the tables to eliminate the need for walking, probably cost absolutely nothing and took less than an hour to implement. Redesign of the job to eliminate the walking doubled output.

Figure 5-1 shows how bricklaying was done before and after Frank Gilbreth introduced the seemingly obvious idea of delivering the bricks at waist level by means of a nonstooping scaffold.

In retrospect, it is clear that the mason could not be paid to lower and raise his entire upper body weight to pick up a five-pound brick. Since the job did get done and the walls were of good quality, though, no one noticed anything wrong with the way they had always done it.

Frederick Winslow Taylor noticed similarly that workers at Bethlehem Steel often used the same shovel for all kinds of material. If they were moving iron ore, the loads were far too heavy, and the workers soon exhausted themselves. If they were moving relatively fluffy rice coal and ash, they

PRODUCTIVITY IMPROVEMENT

"The usual method of providing the bricklayer with material" (Figure 9 in Gilbreth's *Motion Study*. The photo is dated 9/5/1906).

"Nonstooping scaffold designed so that uprights are out of the bricklayer's way whenever reaching for brick and mortar at the same time" (Figure 1 in Gilbreth's *Motion Study*). Notice the bricks' orderly arrangement in packets, and the fact that the bricks and the wall are both at approximately waist level.

Figure 5-1. Bricklaying before and after Gilbreth's improvements.

probably put as much effort into moving the shovel as they did into moving the material. In this case, the joke about using a toothbrush to clean a floor comes to mind.

Gilbreth's work may have resulted in threefold or so improvements, but Henry Ford's achievements show just how much inefficiency can actually be built into work. A contemporary source credited his methods with hundredfold improvements:

> Ford's success has startled the country, almost the world, financially, industrially, mechanically. It exhibits in higher degree than most persons would have thought possible the seemingly contradictory requirements of true efficiency, which are: constant increase of quality, great increase of pay to the workers, repeated reduction in cost to the consumer. And with these appears, as at once cause and effect, an absolutely incredible enlargement of output reaching

something like one hundredfold in less than ten years, and an enormous profit to the manufacturer.[23]

Ford laid out job design principles that are both easy and important to remember: "We now have two general principles in all operations—that a man shall never have to take more than one step, if possibly it can be avoided, and that no man need ever stoop over."[24] Remember that walking and bending add no value to whatever is being done but they do increase its cycle time and time variation.

Rigid work holders (as opposed to overhead conveyors) may also provide some ergonomic or motion efficiency advantages. In many factories, large pieces travel on chains that are attached to overhead conveyors—a practice that may have originated in the meat packing factories that helped inspire the assembly line. A potential disadvantage is that, if the parts can swing or twist, it is harder for people to work on them.

A Toyota ad in a 2005 issue of *U.S. News & World Report* showed a picture of the company's engine production line in West Virginia, and any disciple of the Henry Ford thought process can infer a lot of information from it. The engines travel on cylindrical pillars, and the picture suggests that these pillars may even be able to rotate to present different parts of the engine to the workers. The employees look like they are riding a conveyor belt so they do not have to walk to keep up with the work, and their tasks are between waist and chest level in all cases. It is easy to see why the plant earned the 2005 Harbour Award for the most productive engine plant.[25]

IN-LINE QUALITY CONTROL

Scrap is generally bad and, as shown in Chapter 1, it is even worse in or after the constraint. This is because the constraint, by definition, lacks the surplus capacity to replace the losses. For this reason, it is vital to prevent nonconforming pieces from entering the constraint and wasting its irreplaceable capacity.

Another consequence of scrap, aside from the wasted resources, is its aggravation of the variation that keeps factories from "rolling a six every time." If 1 percent of the pieces are scrap, the factory must make 101 pieces for every hundred that are needed. The problem is that making 101 pieces does not guarantee delivery of 100. Assuming a binomial distribution as the model for scrap, there is better than a 25 percent chance of losing two or more pieces. The factory must then start 103 pieces to be 98 percent sure of delivering 100. This does not sound like a major problem until one thinks of it in the context of a highly complex bill of materials in which a shortfall of any item will reduce the throughput of finished goods.

There is also the problem of being nickel-and-dimed to death by seemingly low scrap rates. One percent does not sound like a high scrap rate, but it becomes devastating if it affects every operation in a long process. *Rolled throughput yield* (RTY) shows just how much damage these seemingly minor losses can cause (see the following equation). For a process with n operations, each of which has a p_i nonconforming fraction, the rolled throughput yield is $RTY = \prod_{i=1}^{n} (1 - p_i)$.

For example, suppose that 20 operations in series each have a 1 percent scrap rate. The RTY is then 81.8 percent; that is, 18.2 percent of the production starts are lost. The problem can be alleviated by parallel processing; that is, two ten-step processes followed by assembly of their outputs will have close to a 90 percent yield under these conditions as well as a shorter cycle time. As shown by Table 5-1, parallel assembly is almost invariably superior to series assembly.

Table 5-1. Rolled throughput yield, parallel and series assembly.

Parallel		Series
$\prod_{i=1}^{n1} (1 - p_i) \rightarrow A$	and then assemble A and B to get finished product C is superior to:	$\prod_{i=1}^{n1+n2} (1 - p_i) \rightarrow C$
$\prod_{i=1}^{n2} (1 - p_i) \rightarrow B$		

The problem is that many products, such as semiconductor devices, cannot be fabricated through parallel processing. Semiconductors are effectively built one layer at a time on silicon wafers, and their microscopic diodes and transistors cannot be assembled by parallel processing steps.

The fact that scrap and rework reduce effective capacity is straightforward, and the term *hidden plant* refers to the additional resources that must be kept on hand to make up for the waste due to poor quality. Hopp and Spearman add that rework also increases cycle time and variation, and that long rework loops make the problem even worse.[26] If the part must go back several operations, it affects the capacity of every workstation that it must use. Rework can become so disruptive that some factories add separate lines for dealing with it. Needless to say, a production line that exists solely to handle rework is a hidden plant and is, therefore, muda (or waste).

Error-Proofing (Poka-Yoke)

It is a basic principle of manufacturing process design that any mistake whose prevention relies on operator vigilance is sure to occur sooner or later. If a job can be done wrong, it eventually will. This is the concept behind error-proofing, or poka-yoke. Error-proofing devices and self-check systems support the lean manufacturing principle: "Don't take it [poor quality], don't make it, and don't pass it on." Self-check systems can be treated as a special class of error-proofing devices that, instead of actually preventing mistakes, catch them before they can cause significant harm.

Error-proofing reflects the idea that even careful workers will eventually make mistakes. For example, if two parts can be assembled backward, it is only a matter of time until this happens unless there are keys and slots that prevent assembly in any but the desired orientation. The use of keys, slots, and other devices was originally called fool-proofing (*baka-yoke* in Japanese), a term to which skilled and industrious Japanese workers rightly took exception.[27] The term was changed to error-proofing to reflect the fact that even diligent workers will eventually make mistakes if the opportunity exists.

The Automotive Industry Action Group provides an example of error-proofing the attachment of an inertia brake to a heavy-duty transmission. The brake has a tang on one corner that will fit the adapter plate only if the orientation is correct. The adapter plate, meanwhile, has a tang that will fit the transmission only in the correct orientation.[28]

Self-Check Systems

Shigeo Shingo refers to self-check systems as *100 percent inspection*, which can be deceptive to quality practitioners who know quite well that trying to inspect quality into a product is the last resort of production systems that can't build it in. Actual inspection is an inefficient use of human resources. It is often ineffective as well, noting the rule of thumb that one in five

PRODUCTIVITY IMPROVEMENT

defects will escape detection. Shingo's "100 percent inspection" is not, however, inspection at all; it involves the use of automatic self-check systems to prevent workstations from taking poor quality or passing it on. If error-proofing prevents parts from being assembled backward, self-check systems prevent nonconforming parts from being assembled into anything at all. Henry Ford described the operation of such a system; the product had to fit through a gage that was set at the upper specification limit for size, but not through one that was set at the lower specification limit, before it was accepted.[29] Every rod had to pass this gauge, and this is what Shigeo Shingo would call 100 percent inspection even though no human inspector is involved.

Shingo also provides a diagram of a self-check system for stem tighteners whose specification is [9.5, 10.5] millimeters.[30] The automatic sorter consists of a chute with diagonal gauges above it. The first's clearance is 10.5 mm, and it deflects parts that are too large into a rejection bin. Parts that are thinner than 10.5 mm continue down the chute to a gauge with 9.5 mm of clearance. Those that are below specification continue through that gauge into another rejection bin, while those that meet specification are deflected into an acceptance bin. The principle is exactly the same as the one for Ford's bushing sorter. Furthermore, any rejection triggers an audible signal (buzzer) to alert the operator to the problem.

Another example from Shingo involves a four-spindle drill whose bits sometimes broke or detached because of their small diameters.[31] This resulted in incompletely drilled holes. Shingo's figure shows a verification unit that checks the part that has just come out of the drill. If any of its four probes cannot go all the way through the part, it pushes a limit switch that shuts down the machine and turns on a flashing red light (andon light, a visual control) to alert the operator. This is an example of the combination of poka-yoke with a self-check system.

It's conceivable that the system could be improved further through autonomation; as an example, it might be possible to detect excessive torque on any drill so the bit will not break or detach in the first place.

Many self-check and error-proofing devices are very inexpensive, often on the order of less than a hundred dollars. Because losses in or after the constraint cannot be replaced, the importance of preventing errors and keeping nonconforming pieces away from the constraint cannot be overemphasized.

Where self-check systems are not practical, the "$N = 2$" inspection technique may be used, but only if the process is stable.[32] This means that, if the first and last pieces of a production run are good, the ones in between also should be good.

Source inspections[33] are even more sophisticated in principle, although not necessarily harder to implement in practice.

Source Inspections

Self-check systems prevent defective parts from leaving a workstation, but source inspections prevent the defects from occurring in the first place. Per Shingo, the error is detected and corrected before it turns into a defect. (In the example of the four-spindle drill, the defect—an undrilled hole—occurs but does not get out of the tool.)

> Source inspections can be described as inspection methods that, rather than stimulating feedback and action in response to defects, are based on the idea of discovering errors in conditions that give rise to defects and performing feedback and action at the error stage so as to keep those errors from turning into defects.[34]

Vertical source inspections focus on upstream processes to prevent nonconforming materials from entering the operation in the first place. They tie in with the SIPOC model because of their focus on suppliers, whether internal or external.

Horizontal source inspections focus on the detection of defect sources within the operation in question. Shingo describes an operation that requires workers to include ten accessories and an instruction manual with each vacuum cleaner in a package.[35] Workers sometimes failed to include all the accessories so the operation was error-proofed as follows. The containers for the accessories were equipped with limit switches that tripped whenever an accessory was removed, and a green light turned on over the container. The instruction manual container was similarly equipped. If the worker did not include all the accessories, the packing carton halted and a buzzer sounded. When the worker added the missing item (as indicated by the absence of a green light over the container in question), the operation was allowed to proceed.

Shingo describes similar error-proofing systems that effectively count the number of parts that the worker has withdrawn for assembly into the product, and notifies the worker if one is missing. That is, the error is effectively caught in the act and corrected before it can become a nonconformance.

Autonomation (*Jidoka*)

Self-check systems and source inspections tie in with the concept of autonomation (*jidoka*), or giving machines the ability to detect abnormal conditions and react to them. The ability of the four-spindle drill to detect a broken bit is a *reactive* form of autonomation; that is, the tool is capable of realizing that it has produced a defective part, and then stopping to prevent more defects. The same applies to Sakichi Toyoda's development of a loom that would stop automatically whenever a thread broke, thus preventing the manufacture of defective cloth.

Proactive autonomation, which means the tool can recognize and correct an abnormal condition before any nonconformances occur, is perhaps the mechanized equivalent of a source inspection. The concepts are roughly interchangeable; in all cases, the basic principle is that the system should prevent defects as opposed to finding them later. Kalpakjian describes proactive autonomation techniques that detect tool wearout before a failure actually occurs:

> One recent development is the acoustic emission technique, which utilizes a piezoelectric transducer attached to a toolholder. The transducer picks up signals, which are acoustic emissions resulting from the stress waves generated during cutting. Experimental studies have shown that the acoustic emission, as measured by the mean root-mean square (RMS) of the signal, increases with increasing wear.[36]

Kalpakjian adds that it is possible to predict the life of a drill or tap by measuring torque as a function of the number of holes drilled or tapped.[37] The torque will be relatively constant until tool wearout begins, whereupon it will start to rise. This allows scheduled replacement of the tools (as preventive maintenance). If the machine can measure its own torque while it is running, it can detect tool wearout automatically.

Summarizing In-line Quality Control

In summary, the following in-line quality control techniques all support the concept, "Don't make it, don't take it, and don't pass it on" with regard to poor quality:

- **Successive check systems** involve the inspection of each piece by an impartial inspector (or by implication an automatic gage or instrument) before it goes into the next workstation. This does not prevent the making of a defect, but it does prevent the next operation from taking it. In contrast to a typical incoming inspection, which is usually applied to a batch or a lot, a successive check system examines each piece immediately and provides immediate information on any defects to the responsible workstation.
- **Self-check systems** move the inspection to the producing workstation. They cannot prevent the making of defects either, but they support the "don't pass it on" principle. In addition, the self-check system's feedback is likely to be more rapid than that of a successive check system. Reactive autonomation, in which the workstation can detect its own production of a nonconformance because of a mechanical failure, is similar in principle.
- **Source inspections** detect and correct potential error sources before they turn into defects. This leads to the ideal "don't make it" state. Proactive autonomation, in which the tool can detect and react to an impending failure before it produces any defects (or damages its own bit or cutter) is an extension of this idea.
- **Poka-yoke** or error-proofing is similar to a source inspection, because it prevents errors from occurring. It is, however, even more proactive. Instead of detecting an error in progress (like omission of a part or assembly of parts in the wrong orientation), it makes the error impossible. Slots, tangs, and keys, for example, make it impossible to assemble parts in any but the correct orientation.

PREVENTIVE MAINTENANCE

Preventive maintenance ties in with the concept of proactive autonomation. It prevents defects or damage to tools by servicing equipment before a problem occurs. Preventive maintenance usually occurs, however, in response to a schedule instead of detection of an impending problem by a machine.

The importance of preventive maintenance in avoiding equipment stoppages cannot be overemphasized. From a capacity standpoint, a stoppage at any nonconstraint is not a serious problem under the Theory of Constraints. Since the activity has excess capacity, it can easily catch up with the delayed work. A balanced system with no excess capacity (Henry Ford's goal) does not have the luxury of allowing any stoppages. Even though a real-world system will undoubtedly have some excess capacity, even a minor stoppage can propagate variation through the entire system. National Semiconductor's position is that there is no such thing as a "minor" stoppage.[38]

A minor stoppage fits the definition of friction, or muda, almost perfectly. Workers can generally fix the problem by themselves, without calling for maintenance or even spare parts. They can do so fairly quickly, and there are no immediate, major consequences of the stoppage. The result is that no one ever bothers to fix the problem's underlying cause, so it keeps coming back.

Edwin P. Norwood shows that Ford's River Rouge plant did not put up with so-called "minor stoppages" either.[39] Workers were actually empowered to stop the line—a practice that most people attribute to Toyota—in the event of trouble. This lit the equivalent of an andon light in a control booth, and the light showed the exact location of the stoppage. If it took more than two minutes for the operator to fix the problem, a maintenance mechanic was called. Even if the operator cleared it within the allotted two minutes, however, the problem and its location were recorded. This allowed a Pareto-type assessment of where stoppages were costing the most time, which, in turn, facilitated many changes and improvements. Modern total productive maintenance (TPM) uses similar practices.

Frederick Winslow Taylor described scheduled preventive maintenance almost 100 years ago. Notices "come out at proper intervals throughout the year for inspection of each element of the system and the inspection and overhauling of all standards as well as the examination and repairs at stated intervals of parts of machines, boilers, engines, belts, etc., likely to wear out or give trouble, thus preventing breakdowns and delays."[40] Today, a good preventive maintenance program is required by ISO 9001:2000 and ISO/TS 16949.

SUPPLY CHAIN MANAGEMENT

> The manufacturer often sees how swiftly the economies in manufacturing are swallowed up in wasteful distribution and this distribution may not be within the control of the manufacturer. There is no point in economizing in manufacturing if at the same time the suppliers and distributors charge all that the traffic will bear.[41]

The role of the supply chain in achieving smooth just-in-time flow—or, if it is poorly designed or managed, accumulating massive inventories that are subject to obsolescence—cannot be overemphasized. The leanest company in the world cannot operate effectively in a batch-and-queue supply chain, or in a supply chain that lacks effective coordination and communications.

Hopp and Spearman cite the *bullwhip effect*[42] in which a complex supply chain can amplify the effects of small demand fluctuations at the retail or end-user piece of the chain. The phenomenon is similar to what can happen in poorly designed feedback control systems that amplify variation by overcorrecting for it.

The Beer Game[43]

The Beer Game was developed at the Massachusetts Institute of Technology in the 1960s. It simulates a supply chain with (1) weeklong lead times between order placement and order delivery, and (2) limited communication between the supply chain partners. The supply chain consists of the retailer, wholesaler, distributor, and manufacturer. Order sheets are the only permitted form of communication, and both backorders and excess inventories incur penalty costs. What usually happens is that random demand fluctuations at the retail end amplify into alternating shortages and overstocks, much like the wave motion in a bullwhip. When the system begins to respond to its own (excessive) adjustments as opposed to genuine fluctuation in the input variable (in this case demand), an uncontrolled runaway situation is almost certain to result. The system is then actually responding to what chemical engineers would recognize as an *unstable feedback loop* as opposed to actual customer demand. Donald R. Coughanowr and Lowell B. Koppel describe situations in which a system's response becomes more oscillatory, and each successive response's amplitude is higher.[44] The accompanying figure in their book could easily be described by the waves of a bullwhip.

What is even more disquieting is the fact that the Beer Game involves a single product and four supply chain partners in series. The effects of a complicated and multilevel bill of materials in a poorly coordinated supply chain are easy to imagine.

Sources of Instability in Supply Chains

Hopp and Spearman cite four major reasons for unstable inventory levels and stockouts in supply chains:[45]

- **Batching of orders to reduce the cost of replenishment orders can amplify relatively small demand fluctuations at the retail level.** This is especially true when everyone places their respective replenishment orders at the same time, such as at the beginning or end of the month. Another motive for order batching is to fill trucks and reduce the per-unit shipment cost. The same source recommends the following measures:
    - Use electronic data interchange (EDI) to reduce the cost of placing purchase orders. It is quite likely that retailers are already using some form of electronic kanban to signal their upstream suppliers in real-time that a certain number of units have been sold and that replacements will be needed.
    - Consolidate orders to fill trucks. *Truck sharing* is a well-known logistics technique in which orders from and for different firms share a truck, like passengers in a bus.
- **Demand forecasting can amplify variability by encouraging order spikes that exceed demand spikes**. As an example, a wholesaler receives an order from a retailer, adds safety stock to the order, and places an even larger order to the upstream distributor. Each level of the supply chain responds to demand from the customer immediately downstream, as opposed to end customer demand. Hopp and Spearman suggest the following measures:
    - *Sell-through data* means that the end seller shares sales data with its upstream supply chain partners.
    - *Vendor-managed inventory* (VMI) means that the manufacturer controls inventory throughout the entire supply chain. (Bill Walker refers to the channel master, or the trading partner who orchestrates the supply chain's operations.[46])

- *Shorten feedback loops.* Long ones make any kind of system harder to control, and supply chains are no exception. Shorter lead times, which may be achieved through lean manufacturing techniques and pull production control, reduce the need to carry safety stock, a principle that Henry Ford described quite explicitly in *My Life and Work*.
- **Pricing policies can drive dysfunctional (from the supply chain's perspective) purchasing decisions.** As an example, knowledgeable customers can game the system when they know that the seller will have to offer a discount to move the inventory. Although both Henry Ford and Benjamin Franklin warned against buying things merely because they are offered at a discount, purchasers can safely ignore this advice when they know the items will be used. Christmas cards, for example, are marked down 40 percent or more on December 26. If one plans to send holiday cards the following year and is willing to keep them around for a year, this is an excellent bargain. Along the same lines, many consumers know that fall is the best time to buy summer clothing and spring the best time to buy winter clothing, often at discounts exceeding 50 percent. Even nonperishable groceries (like cereal) fit this paradigm; they are likely to sit unsold if customers know that a store offers periodic sales. When the sale is announced, they fly off the shelves.
    - Hopp and Spearman suggest "everyday low pricing," which means that the seller maintains a consistent low price somewhere between full retail and discounted sale.
    - Make-to-order as opposed to make-to-forecast also eliminates this problem. If cars were made only to order, customers could not game the system by waiting until the end of the model year and the huge discounts and incentives that everyone knows go with it. On the other hand, elimination of the inventory and its carrying costs would allow lower everyday prices. Books can be printed to order—the technique is known as print-on-demand—and the same could conceivably be done with holiday greeting cards. Cafe-Press can, in fact, print custom-designed greeting cards on demand. The fact that the computer punch card originated to control the Jacquard loom suggests that clothing can be made to order as well.

- **Dysfunctional customer behavior includes placing unrealistically large orders to ensure their supply during expected shortages.** The customers then cancel their orders for any excess. Hopp and Spearman cite a common practice in the 1980s: Customers placed the same order for computer memory chips with multiple suppliers, and then canceled all but the first order to be delivered. Shortening the supply chain is an excellent way to reduce instability as well as costs. Remember that anything that does not add value—value being defined as something for which the customer is willing to pay—is waste. A car dealer's inventory carrying costs, advertising costs, overhead, and markup add no value for the customer, and this suggests that the dealers should be cut out of the supply chain. More precisely, they should be relegated to taking delivery of made-to-order vehicles, along with vehicle service. Dell Computer, meanwhile, cut retailers out of its supply chain and sold its products directly to customers. Customers configure their computers on the Internet, and the computers are then assembled-to-order from readily available parts.

Logistics

Freight management systems and similar logistics systems are absolutely vital to a lean supply chain. Norwood creates a vivid mental picture of the ideal condition: *a continent-spanning conveyor,* which Henry Ford created without the benefit of the Internet or even primitive computers:

> In a word, management has here resolved America's transportation facilities into yet another network of conveyor lines. Using that multitude of additional links offered by rail, highway, water, and air, it has butt-welded them with their own timetables and picketed them with telegraphic checkings as watchful as the straw bosses who supervise progression along the conveyor lines of the shop. There is little of the fanciful in likening the plant conveyor system to that made available by exterior transportation.[47]

The same source adds that the transportation system operated on a just-in-time basis. Ford's transportation department could determine any rail car's location to within 25 miles and estimate its time of arrival with only minutes of uncertainty.[48] This was, of course, long before anyone had heard of the Global Positioning System.

SUMMARY: PRODUCTIVITY IMPROVEMENT

It is impossible to overemphasize the importance of seeing waste, which is also known as friction or muda. No one wants to waste time, money, or materials, but waste frequently hides in plain view. One of Henry Ford's principal success secrets was an ability to recognize waste on sight and to teach this skill to his workforce. False economies, whether in hiring cheap workers or purchasing cheap equipment and materials, can cause enormous waste. The operative phrase is "penny wise and pound foolish," because saving a penny through false economy up front may cost a dollar (or the opportunity to earn a dollar) downstream.

Effective in-line quality control is central to reducing the cost of poor quality while identifying its root causes to prevent recurrence.

- **Error-proofing** makes it impossible to do the job the wrong way. Parts cannot be assembled in the wrong orientation because keys, slots, and similar features allow assembly only in the correct orientation.
- **Sequence inspections** prevent the acceptance of nonconforming work by the next operation in the series while notifying the operation that made the defect.
- **Self-check systems** move the quality assessment to the producing workstation. As an example, parts that fail to pass through a gauge are rejected immediately, and the operator is notified immediately.
- **Source inspections** improve on self-check systems by preventing errors and defects from occurring.
- **Autonomation** (jidoka) means that equipment can detect abnormal conditions and shut down or call for maintenance before it produces defective work.
- **Reactive autonomation** means that the machine can detect a condition like a broken tool, shut itself down, and call for maintenance.
- **Proactive autonomation** means that the machine can determine that a tool is wearing out and call for a replacement before the wear gets bad enough to compromise quality or result in breakage.

PRODUCTIVITY IMPROVEMENT

- **Preventive maintenance** is similar to proactive autonomation because it prevents breakdowns and malfunctions, but it takes place in response to a schedule as opposed to feedback from the tool.

Supply chain inefficiencies can easily undo the efforts of the leanest companies. The importance of communication with downstream middlemen cannot be overemphasized, lest consumer-end variation result in an unstable feedback loop as simulated by the Beer Game. Even better, non-value-adding downstream middlemen should simply be cut out of the supply chain.

Logistics is also a critical supply chain consideration. Henry Ford created a continent-spanning conveyor to move materials to and from his factories, and this is a good intellectual concept. Today, the continent-spanning conveyor may be realized by freight management systems and third-party logistics systems.

Endnotes

1. Clausewitz, *On War*, p. 119.
2. Halpin, *Zero Defects*, pp. 80–81.
3. Peters, *Thriving on Chaos*, p. 323.
4. Schonberger, *World Class Manufacturing*, pp. 14–15.
5. Ford, *Moving Forward*, p. 187.
6. Bennett, *Ford: We Never Called Him Henry*, pp. 32–33.
7. Ford, *My Life and Work*, p. 15.
8. The System Company, *How Scientific Management is Applied*, pp. 88–89.
9. Aitken, *Scientific Management in Action:*.
10. Ford, *My Life and Work*, p. 77.
11. Ford, *Moving Forward*, p. 53.
12. Ford, *Today and Tomorrow*, pp. 119–121.
13. Basset, *When the Workmen Help You Manage*, p. 64.
14. Franklin, whom Henry Ford cited as an influence on his thinking, is arguably the intellectual father of what we now call lean manufacturing.
15. Hardy, *Longbow*, p. 68.
16. The System Company, *How to Get More Out of Your Factory*, p. 106.
17. Ibid., pp.108-111.
18. Gilbreth, *Motion Study*.

19. Cronley, "Curtis E. LeMay: The Enduring 'Big Bomber Man.'"
20. Chen, "States Boost Speed Limits On Major Highways," *Wall Street Journal*, p. D1.
21. Ford, *Today and Tomorrow*, p. 103.
22. The System Company, *How Scientific Management is Applied*, p. 41.
23. Charles Buxton Going, preface to *Ford Methods and the Ford Shops*, by Arnold and Faurote.
24. Ford, *My Life and Work*, p. 80.
25. http://www.prnewswire.com/cgi-bin/stories.pl?ACCT=104&STORY=www/story/09-28-2005/0004133726&EDATE.
26. Hopp and Spearman, *Factory Physics*, pp. 394–395.
27. Shingo, *Zero Quality Control*, p. 45.
28. AIAG, *CQI-10: Effective Problem Solving, A Guideline*, p. 118.
29. Ford, *Moving Forward*, p. 212.
30. Shingo, *Zero Quality Control: Source Inspection and the Poka-Yoke System*, p. 79.
31. Ibid., pp. 80–81.
32. Schonberger, *Japanese Manufacturing Techniques: Nine Hidden Lessons in Simplicity*, p. 71.
33. Shingo, *Zero Quality Control*, pp. 82–87.
34. Robinson, ed., *Modern Approaches to Manufacturing Improvement*, p. 280.
35. Shingo, *Zero Quality Control*, pp. 86–87.
36. Kalpakjian, *Manufacturing Processes for Engineering Materials*, p. 498.
37. Ibid., pp. 509–510.
38. Gardner and Nappi, "The Total Impact of Minor Stoppages."
39. Norwood, *Ford: Men and Methods*, pp. 10–12.
40. Taylor, *Shop Management*, p. 117.
41. Ford, *Moving Forward*, p. 18.
42. Hopp and Spearman, *Factory Physics*, pp. 612–616.
43. http://beergame.mit.edu/guide.htm.
44. Coughanowr and Koppel, *Process Systems Analysis and Control*, pp. 150–151.
45. Hopp and Spearman, *Factory Physics*, pp. 612–615.
46. Walker, "Supply Chain Management."
47. Norwood, *Ford: Men and Methods*, p. 21.
48. Ibid, p. 30.

APPENDIX

LINEAR PROGRAMMING IN MICROSOFT EXCEL

Microsoft Excel's Solver tool is capable of solving linear programming problems to help identify factory constraints and select optimum product mixtures. Revisit the example from Chapter 1 (presented in the section entitled **Simplex Method**) in which the objective was to (1) select the product mixture that maximizes the marginal profit and (2) identify the constraints. The marginal profits and constraints appear in Table A-1.

Table A-1. Product marginal profits and resource requirements.

Product	A	B	C	
Marginal profit per machine	$6	$6	$4	Available time in minutes
Drilling	$a_{11} = 4$	$a_{12} = 8$	$a_{13} = 6$	480
Grinding	$a_{21} = 6$	$a_{22} = 4$	$a_{23} = 4$	480
Heat treatment	$a_{31} = 8$	$a_{32} = 6$	$a_{33} = 10$	480

As the table shows, product A earns a marginal profit of $6 and requires four minutes on the drill press, six minutes of grinding, and eight minutes of heat treatment. An Excel spreadsheet for solving this problem appears in Figure A-1.

	A	B	C	D	E	F
1	Linear Programming: Simplex Method					
2				Product A	Product B	Product C
3	Marginal Profit	0	Number to produce	0	0	0
4	Constraint	Available	Marginal Profit==> Amount used	6	6	4
5	Drilling	480	0	4	8	6
6	Grinding	480	0	6	4	4
7	Heat Treatment	480	0	8	6	10

Figure A-1. Spreadsheet setup for a simple linear programming problem.

APPENDIX

Table A-2. Cell formulas for linear programming in Excel.

B3 (marginal profit)	{ =SUM(D3:F3*D4:F4)}
	The bracketed array formula is entered by typing in "=SUM(D3:F3*D4:F4)" and pressing Control-Shift-Enter. It computes D3 × D4 + E3 × E4 + F3 × F4 and is equivalent to $\sum_{i=1}^{n} x_i P_i$ where x_i is the production quantity of the *i*th product (A, B, and C in this case) and P_i is its marginal profit.
B5 (drilling time used)	{=SUM(($D$3:$F$3)*(D5:F5))}
	As with the marginal profit entry, this array formula computes the sum of the minutes of drilling capacity that the solution consumes. It is equivalent to $\sum_{i=1}^{n} x_i a_{1i}$.
B6 (grinding time used)	{=SUM(($D$3:$F$3)*(D6:F6))} = $\sum_{i=1}^{n} x_i a_{2i}$
B7 (heat treatment time)	{=SUM(($D$3:$F$3)*(D7:F7))} = $\sum_{i=1}^{n} x_i a_{3i}$

Microsoft Office comes with SOLVSAMP.XLS, which shows how to use Solver. The Product Mix sheet has an example of a product mixture optimization problem. In Figure A-1, cell B3 is the *target cell* that is to be maximized, minimized, or set equal to a specific quantity. Cells D3 through F3 are the *changing cells*, or variables that are to be manipulated to solve the problem. Cells B5 through C7 are the *constraints*. Figure A-2 shows how to use the Solver tool. Note that, if Solver does not appear in the tools menu, it is necessary to install it as an add-in (see add-ins under the tools menu bar).

APPENDIX

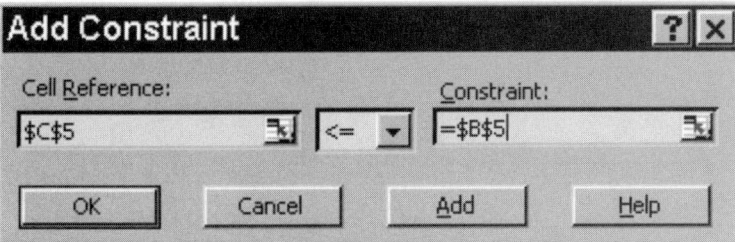

To add constraints, click the "Add" button.

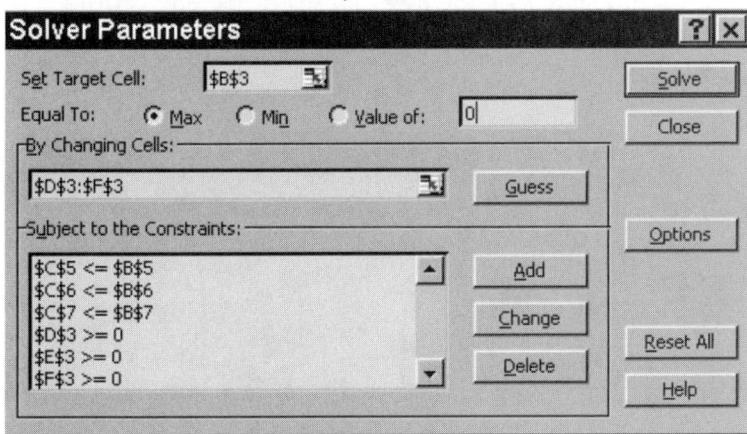

After addition of all necessary constraints:

Figure A-2. The Solver interface.

APPENDIX

Note that, although it is generally taken for granted that a linear programming problem's variables will be nonnegative, the ">=0" constraints must be included. This may, however, be streamlined somewhat, as shown in Figure A-3.

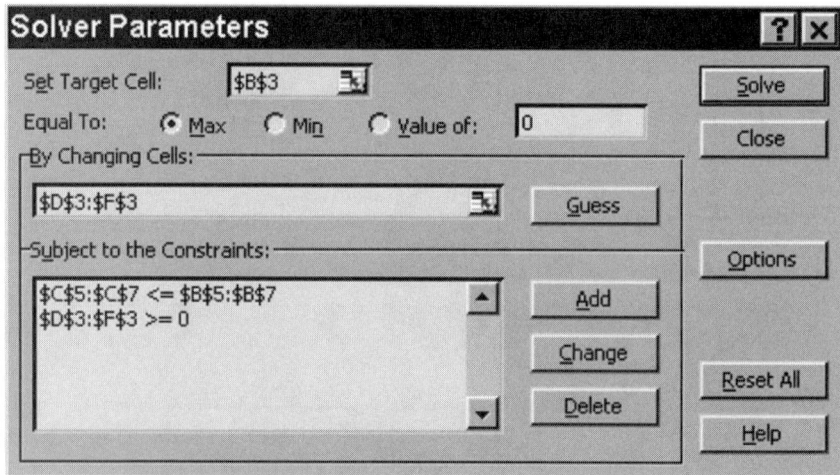

Figure A-3. Entry of constraints by cell ranges instead of individually.

As shown in Figure A-4, the user may select the reports that are to be displayed. Each report will appear in a new worksheet with its own tab in the Excel workbook.

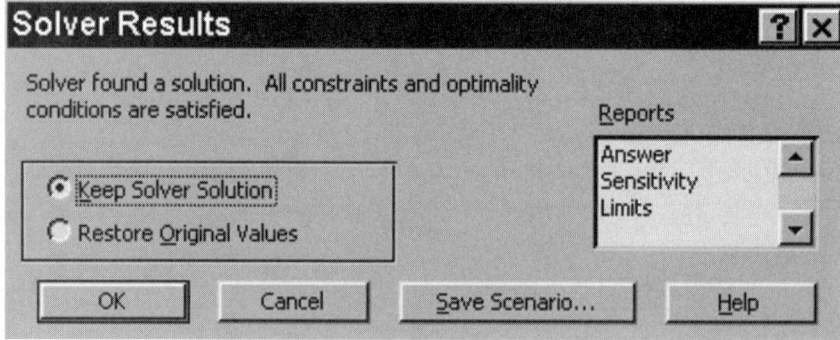

Figure A-4. Available Solver reports.

APPENDIX

In addition, the user should tell Solver to assume a linear model as shown in Figure A-5. It will actually work (at least for a simple problem) with what is apparently the Lagrange multiplier optimization method for nonlinear models, but it is best to use the standard method.

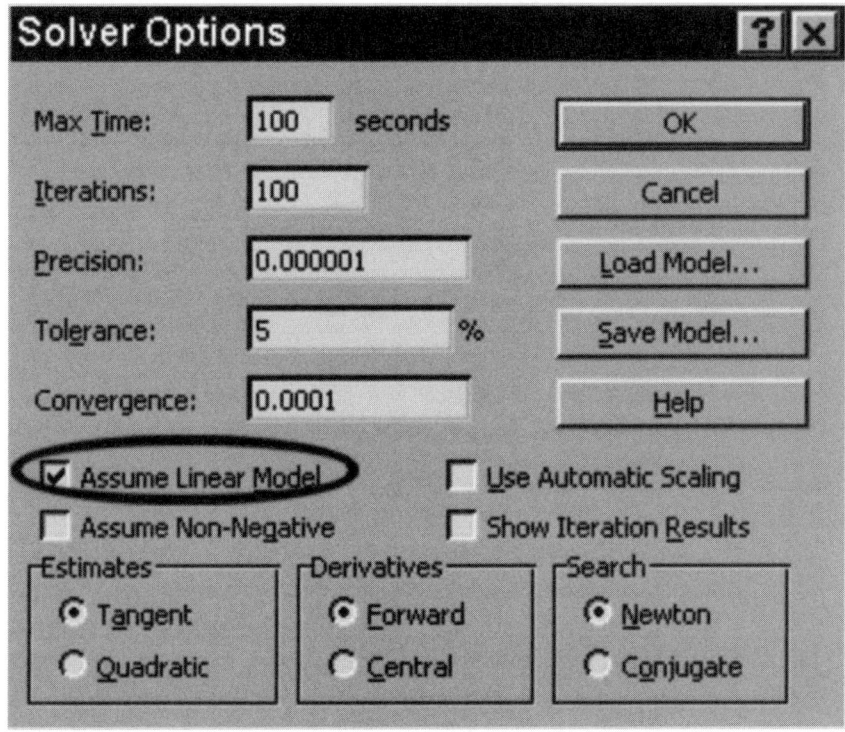

Figure A-5. Options Menu.

Figure A-6 shows the *answer report*, which contains (1) the marginal profit from the optimum product mixture, (2) the product mixture that achieves this, and (3) the slack variables for each constraint. Capacity-constraining resources are "binding" and those with excess capacity are "nonbinding." For practical purposes, the standard ">=0" constraints may be ignored. The sensitivity report (Figure A-7) shows the shadow prices.

APPENDIX

Microsoft Excel 10.0 Answer Report
Worksheet: [LPSolver.xls]SimpleConstraint
Report Created: 8/18/2006 6:04:01 PM

Target Cell (Max)

Cell	Name	Original Value	Final Value
$B$3	Marginal Profit	0	432

Adjustable Cells

Cell	Name	Original Value	Final Value
$D$3	Number to produce Product A	0	24
$E$3	Number to produce Product B	0	48
$F$3	Number to produce Product C	0	0

Constraints

Cell	Name	Cell Value	Formula	Status	Slack
$C$5	Drilling Marginal Profit==> Amount used	480	$C$5<=$B$5	Binding	0
$C$6	Grinding Marginal Profit==> Amount used	336	$C$6<=$B$6	Not Binding	144
$C$7	Heat Treatment Marginal Profit==> Amount used	480	$C$7<=$B$7	Binding	0
$D$3	Number to produce Product A	24	$D$3>=0	Not Binding	24
$E$3	Number to produce Product B	48	$E$3>=0	Not Binding	48
$F$3	Number to produce Product C	0	$F$3>=0	Binding	0

Figure A-6. Solver Answer Report with solutions and constraints.

Microsoft Excel 10.0 Sensitivity Report
Worksheet: [LPSolver.xls]SimpleConstraint
Report Created: 8/18/2006 6:04:01 PM

Adjustable Cells

Cell	Name	Final Value	Reduced Cost	Objective Coefficient	Allowable Increase	Allowable Decrease
$D$3	Number to produce Product A	24	0	6	2	3
$E$3	Number to produce Product B	48	0	6	6	1.5
$F$3	Number to produce Product C	0	-3.8	4	3.8	1E+30

Constraints

Cell	Name	Final Value	Shadow Price	Constraint R.H. Side	Allowable Increase	Allowable Decrease
$C$5	Drilling Marginal Profit==> Amount used	480	0.3	480	160	240
$C$6	Grinding Marginal Profit==> Amount used	336	0	480	1E+30	144
$C$7	Heat Treatment Marginal Profit==> Amount used	480	0.6	480	180	120

Figure A-7. Solver Sensitivity Report.

The allowable increase and decrease are the amounts by which the resource can be changed before the shadow price no longer applies. In other words, the shadow price of 30 cents applies to minutes of drilling in the range [240,640] minutes. The 60 cent shadow price applies to minutes of heat treatment in the range [360,660] minutes.

APPENDIX

Grinding minutes are not a constraint; hence the infinite allowable increase. The allowable decrease is also self-explanatory; it equals the slack for this resource. Taking away up to 144 minutes of grinding time will not change the shadow price of zero, but taking away more will turn grinding into a constraint.

BIBLIOGRAPHY

Aitken, Hugh G. J. 1985. *Scientific Management in Action: Taylorism at Watertown Arsenal, 1908–1915.* Princeton, NJ: Princeton University Press.

Anthony, Robert, and James S. Reece. 1983. *Accounting: Text and Cases*, 7th ed. Homewood, IL: Richard D. Irwin.

Arnold, Horace Lucien, and Fay Leone Faurote. 1915. "Ford Methods and the Ford Shops." New York: *The Engineering Magazine*. Reprinted 1998, North Stratford, NH: Ayer Company Publishers.

Automotive Industry Action Group (AIAG). 2006. *CQI-10: Effective Problem Solving, A Guideline.* Southfield, MI.

Basset, William. 1919. *When the Workmen Help You Manage.* New York: The Century Company.

Begley, Sharon. 2005. "How Brief Drop in Cars Can Trigger Tie-Ups, and Other Traffic Tales." *Wall Street Journal*, Science Journal section, July 1, 2005.

Bennett, Harry, as told to Paul Marcus. 1951. *Ford: We Never Called Him Henry.* New York: Tom Doherty Associates.

Chen, Stephanie. 2006. "States Boost Speed Limits On Major Highways." *Wall Street Journal,* July 20, 2006; page D1.

Clausewitz, Carl von. 1976. *On War.* Translated by M. Howard and P. Paret. Princeton, NJ: Princeton University Press.

Coughanowr, Donald R., and Lowell B. Koppel. 1965. *Process Systems Analysis and Control.* New York: McGraw-Hill.

Cronley, Major T. J. 1986. "Curtis E. LeMay: The Enduring 'Big Bomber Man.'" United States Marine Corps Command and Staff College, Quantico, Va. (www.globalsecurity.org/wmd/library/report/1986/CTJ.htm).

Cubberly, William H., and Ramon Bakerjian, eds. 1989. *Tool and Manufacturing Engineers Handbook*, Desk Edition. Dearborn, MI: Society of Manufacturing Engineers.

Dana, Richard Henry. 1909–1914. *Two Years Before the Mast.* New York: P.F. Collier & Son.

Ford, Henry, and Samuel Crowther. 1930. *Moving Forward.* New York: Doubleday, Doran, & Company.

Ford, Henry, and Samuel Crowther. 1922. *My Life and Work*. New York: Doubleday, Page & Company.

Ford, Henry, and Samuel Crowther. 1926. *Today and Tomorrow*. New York: Doubleday, Page & Company. (Reprint available from Productivity Press, 1988.)

Franklin, Benjamin. 1758 and 1986. *The Way to Wealth*. Bedford, MA: Applewood Books.

Franklin, Benjamin. 1733 and 2002. *Poor Richard's Almanack*. Bedford, MA: Applewood Books.

Gardner, Les, and Frank Nappi. 2001. "The Total Impact of Minor Stoppages." The 6th Annual Lean Management and TPM Conference, sponsored by Productivity Inc. October 25–26, 2001, Dearborn, MI.

Gilbreth, Frank. 1911. *Motion Study*. New York: D. Van Nostrand Reinhold.

Goldratt, Eliyahu, and Jeff Cox. 1992. *The Goal*. 2nd rev. ed. Croton-on-Hudson, NY: North River Press.

Gourley, Catherine. 1997. *Wheels of Time: A Biography of Henry Ford*. Brookfield, CT: The Millbrook Press.

Halpin, J. F. 1966. *Zero Defects*. New York: McGraw-Hill.

Hardy, Robert. 1992. *Longbow*. Bois d'Arc Press, New York: Lyons & Burford.

Harrington, H. James. 2006. "Looks Good on Paper: Inflated accounting systems cost dearly in inventory, scrap, and rework." *Quality Digest*, July 2006 (www.qualitydigest.com/july06/columnists/jharrington.shtml).

Hayes, Robert H., Steven C. Wheelwright, and Kim B. Clark. 1988. *Dynamic Manufacturing: Creating the Learning Organization*. New York: The Free Press.

Heizer, Jay, and Barry Render. 1991. *Production and Operations Management*. 2nd ed. Needham Heights, MA: Allyn and Bacon.

Hillier, Frederick S., and Gerald J. Lieberman. 1980. *Introduction to Operations Research*. Oakland, CA: Holden-Day.

Hogg, Ian V. 1983. *Guns and How They Work*. Secaucus, NJ: Chartwell Books.

Hopp, Wallace J., and Mark L. Spearman. 2000. *Factory Physics*. 2nd ed. New York: McGraw-Hill.

Hoyer, R. W. 2001. "Why Quality Gets an 'F': What happens when the bottom line overrides a focus on customer needs." *Quality Progress*, October 2001, 32–36.

BIBLIOGRAPHY

Hradesky, John. 1986. *Productivity & Quality Improvement: A Practical Guide to Implementing Statistical Process Control.* New York: McGraw-Hill.

Imai, Masaaki. 1997. *Gemba Kaizen.* New York: McGraw-Hill.

Kalpakjian, Serope. 1984. *Manufacturing Processes for Engineering Materials.* Reading, MA: Addison-Wesley.

Kipling, Rudyard. 1897. *Captains Courageous.* New York: Century.

Law, Averill M., and W. David Kelton. 1982. *Simulation, Modeling, and Analysis.* New York: McGraw-Hill.

Levinson, William. 2002. *Henry Ford's Lean Vision: Enduring Principles from the First Ford Motor Plants.* Portland, OR: Productivity Press.

Levinson, William, and Raymond Rerick. 2002. *Lean Enterprise: A Synergistic Approach to Minimizing Waste.* Milwaukee, WI: ASQ Quality Press.

Melville, Herman. 1849 and 2004. *Redburn: His First Voyage.* Whitefish, MT: Kessinger Publishing.

Menning, Bruce W. 1986. "Train Hard, Fight Easy: The Legacy of A. V. Suvorov and His 'Art of Victory.'" *Air University Review*, December 1986, 79–88.

Norwood, Edwin P. 1931. *Ford: Men and Methods.* Garden City, NY: Doubleday, Doran & Company.

Ohno, Taiichi. 1988. *Toyota Production System.* Portland, OR: Productivity Press.

Parkhurst, Frederic A. 1912. *Applied Methods of Scientific Management.* Easton: Hive Publishing Company (1980 reprint published by Wiley, NY).

Peters, Tom 1987. *Thriving on Chaos.* New York: Harper & Row.

Preston, Anthony. 1981. *Battleships.* Bison Books Ltd.

Quality Council of Indiana, 2001. *Certified Six Sigma Black Belt Primer.* West Terre Haute, IN.

Rigg, James L. 1977. *Engineering Economics.* New York: McGraw-Hill.

Robinson, Alan, ed. 1990. *Modern Approaches to Manufacturing Improvement: The Shingo System.* Portland: Productivity Press.

Sawyer, Christopher. 2001. "Hot off the Skillet." *Automotive Design and Production*, February 2001. (http://www.autofieldguide.com/columns/0201oncar.html)

Schonberger, Richard J. 1982. *Japanese Manufacturing Techniques: Nine Hidden Lessons in Simplicity.* New York: The Free Press.

BIBLIOGRAPHY

Schonberger, Richard J. 1986. *World Class Manufacturing.* New York: The Free Press.

Seo, K. K. 1984. *Managerial Economics: Text, Problems, and Short Cases.* Homewood IL: Richard D. Irwin.

Sheckley, Robert. 1968. "The Laxian Key" in *The People Trap.* New York: Dell Publishing.

Shingo, Shigeo (Andrew Dillon, trans.). 1987. *The Sayings of Shigeo Shingo: Key Strategies for Plant Improvement.* Portland, OR: Productivity Press.

Shingo, Shigeo. 1986. *Zero Quality Control: Source Inspection and the Poka-Yoke System.* Portland, OR: Productivity Press.

Smith, Wayne. 1998. *Time Out: Using Visible Pull Systems to Drive Process Improvement.* New York: John Wiley & Sons.

Sorensen, Charles E., with Samuel T. Williamson. 1956. *My Forty Years with Ford.* New York: W. W. Norton.

Standard, Charles, and Dale Davis. 1999. *Running Today's Factory: A Proven Strategy for Lean Manufacturing.* Cincinnati: Hanser Gardner Publications.

Suzaki, Kiyoshi. 1987. *The New Manufacturing Challenge: Techniques for Continuous Improvement.* New York: The Free Press.

Taylor, Frederick Winslow. 1911. *Shop Management.* New York: Harper & Brothers Publishers.

The System Company. 1911. *How Scientific Management is Applied.* London: A. W. Shaw Company Ltd.

The System Company. 1911. *How to Get More Out of Your Factory.* London: A. W. Shaw Company, Ltd.

Turbide, Dave. 2004. "A Lean Approach to Lean." *APICS—The Performance Advantage,* July/August 2004.

Voiland, Douglas E. 2001. "A Nice Problem to Have: A commonsense approach to TOC can save even the most successful company." *APICS—The Performance Advantage.* July 2001, 29–31.

Von Steuben, Friedrich. 1774. *Regulations for the Order and Discipline of the Troops of the United States.* Philadelphia: Styner and Cist

Walker, Bill. 2001. "Supply Chain Management," APICS-NEPA meeting, Pittston, PA, March 14, 2001.

Womack, James P., and Daniel T. Jones. 1996. *Lean Thinking.* New York: Simon & Schuster.

INDEX

AB:POM, 25, 26, 27
Accounting
   cost, 111
   cycle time, 80–82
   time, 80, 103
AIAG. *See* Automotive Industry Action Group
Andon lights, 39
Arnold, Horace Lucien, 17
Arquebus, 113
Artillery breech blocks, 95
Assembly line, moving, 102, 104
Automotive Industry Action Group (AIAG), 78, 79, 103
   on error-proofing, 122
Autonomation, 125, 132
   proactive, 125, 132
   reactive, 125, 132

B-24 Liberator bomber, 36
Baker Rifle, 93
Bar coding, 80
Basic variable, 21
   entering, 22
   leaving, 22
Batch reactors, 50
Batching, 70, 87, 104
   alternatives to, 88–90
   disadvantages of, 85
   encouraging, 97
   Ford on, 87
   process, 60, 84–85, 87
   to reduce cost, 129
   transfer, 60–61, 84–85, 86
Beer Game, 128–29, 133
Begley, Sharon, 66

Benetton, 11
Bennett, Harry, 108–9
Bethlehem Steel, 110, 118
Beyond the Theory of Constraints, 100
Bill of materials (BOM), 16, 35, 59
BOM. *See* Bill of materials
Boy Scout hike, 1
Breech-loading artillery, 95
Bricklaying, 118, 119
Brown, Burke, 52
Brown, G. Donaldson, 8
Buffer, 2
   inventory, 42, 43
   management, 43, 45
Bullwhip effect, 128

CafePress, 11, 130
Capacity-constraining resource (CCR), 1, 3, 21, 29
   feast-or-famine situation at, 43
   stoppage at, 43
Captains Courageous (Kipling), 73, 76, 77, 100
Cart sharing, 86
CCR. *See* Capacity-constraining resource
Cellular manufacturing, 98
Chaplin, Charlie, 73
Chemical process industry (CPI), 49
Christmas cards, 130
A Christmas Carol (Dickens), 115
Clark, Kim B., 36
Clausewitz, Carl von, 107
Constraint, 29
   analytical identification of, 18–19
   breaking setup at, 92

INDEX

discovery of, 18
elevating, 77
factory, 135
market, 27–28
in Microsoft Excel, 136, 138
rework in, 6
stoppage in, 5
Container ship, 115
Continental Army, 113
Continent-spanning conveyor, 131, 133
Continuous stirred tank reactor (CSTR), 50
Continuous-flow chemical process equipment, 88
Contractual requirements, 28–29
Convair, 36
CONWIP, 46–47
Cost
   accounting, 111
   batching to reduce, 129
   differential, 15, 29
   of home ownership, 4
   labor, 5
   marginal, 15, 30
   material, 5
   opportunity, 3–5, 29, 30
   overhead, 5
   rework, 5
   Rigg on, 4
   of spaghetti diagram, 97
   sunk, 12–14, 17, 30
   time as, 3
   TOC and variable, 15–17
   transportation, 9
   variable, 16
Coughanowr, Donald R., 128
Countermarching, 113
Cox, Jeff, ix, 1
CPI. *See* Chemical process industry
CPM. *See* Critical Path Method
Cratchit, Bob, 115

Critical Chain, 44
Critical Path Method (CPM), 44
Critical ratio prioritization rule, 46
Crossbow, 112
CSTR. *See* Continuous stirred tank reactor
Customer behavior, 131
Cycle time, 3, 4, 7
   accounting, 80–82
   equation for, 68
   increase in, 85
   measuring, 103
   of single-server queue, 67

DaimlerChrysler, 52
Dana, Richard Henry, Jr., 37–38
Davis, Dale, 8, 9
   on pig-swallowing, 33–34
DBR. *See* Drum-buffer-rope system
Decision, 79
Dedicated equipment by product family, 103
Delay, 79, 80, 103
Dell Computer, 131
Demand forecasting, 129
Department of Energy, 117
Departmental layout, 102
Dickens, Charles, 115
Differential cash flow, 6
"Don't take it, don't make it, don't pass it on," 61, 122, 126
Drum-buffer-rope system (DBR), x, 33, 42–49, 43, 50, 54, 77
   characteristics of, 42

Economic order quantity (EOQ), 90
EDI. *See* Electronic data interchange
Electronic data interchange (EDI), 129
Environmental and Energy Study Institute, 117
EOQ. *See* Economic order quantity

INDEX

Equipment
  cheap, 110–11
  dedicated, 103
  utilization, 7, 11–12
Error-proofing, 122, 126, 132
"Everyday low pricing," 130

False economy, 109–17, 132
  cause-and-effect diagram for, 110
  of cheap equipment, 110–11
  of cheap labor, 111–14
  of cheap procedures, 116–17
  of cheap purchases, 114–15
  of cheap working conditions, 115–16
The Family Circus, 97
Farming, 109
Faroute, Fay Leone, 17
"First-class man," 113, 114
Fish-cleaning operation, 73, 75, 100
  subdivided, 76
"Five Whys" technique, 82
  examples of, 82–83
  Ford and, 83
  key point of, 83
Flow, 49–53
  at start of 20th century, 52–53
Flow lines, 50, 54
Fluorescent light, 115
Flying boxes, 52–53
FMS. *See* Freight management system
Fool-proofing, 122
Ford, Henry, ix, 3, 10
  on batching, 87
  on buying, 10, 130
  on cheap labor, 111–14
  complaints against, 73
  "Five Whys" technique and, 83
  on friction, 108
  goal of, 127
  individual electric motor introduced by, 99
  on JIT, 34–35

  job design principles of, 74, 120
  lean manufacturing and, 8
  principles of assembly by, 98
  on sunk costs, 12, 17
  on wages, 112
  on waste, 108–9, 132
Ford Motor Company
  axle manufacturing at, 78
  Dearborn steel mills of, 10
  Highland Park factory of, 18, 89, 98
  JIT at, 34–35
  River Rouge plant of, 51, 127
  self-check systems at, 61
  single-unit flow and, 89
Four-spindle drill, 123, 125
Franklin, Benjamin, 3
  on buying, 10, 130
  on dysfunctional purchasing, 9
  on longbow, 113, 114
Free Producer, 14
Freight management system (FMS), 35, 131
Friction, 107, 127
  Ford on, 108
  Halpin on, 108
  key characteristics of, 108
  Peters on, 108
  Schonberger on, 108

Gantt chart, 81, 98
Gantt, H. L., 98
Gatling gun, 101
Gilbreth, Frank, 75, 79, 92
Global Positioning System (GPS), 132
The Goal (Goldratt and Cox), ix, 1
Goldratt, Eliyahu, ix, 1, 44
Golf, 91
  purpose of, 92
Good money after bad, 13
GPS. *See* Global Positioning System
Greek phalanx, 37
Grinding minutes, 141

INDEX

Group technology, 100
Gunpowder, 93
  introduction of, 37, 95

Halpin, J. F., 108
Handle, 80, 103
Handling, 101
Harrington, James, 8
Hayes, Robert H., 36
Heat treatment capacity, 23, 100, 140
Heizer, Jay, 25
Henry Ford's Lean Vision, 75
Hidden plant, 121
Hilker and Wiechers Company, 53
Home ownership, cost of, 4
Hopp, Wallace, 11
Hradesky, John, 78
Human motion, 79

Imai, Masaaki, 52
Industrial statistics, 75
In-line quality control, 120–26, 132
  error-proofing as, 122, 126, 132
  self-check systems as, 122–24, 126, 132
  sequence inspections as, 132
  source inspections as, 124–25, 126, 132
  successive check systems as, 126
  summary of, 126
Inspection, 79, 80, 103
Internal Revenue Service, 11
Inventory, 7, 29
ISO 9001:2000, 127
ISO/TS 16949, 127

Jacobs, W. W., 6
Jacquard loom, 130
Jidoka. *See* Autonomation
Jinking, 116
JIT. *See* "Just-in-time"

"Just-in-time" (JIT), x
  delivery, 46
  key aspects of, 35
  manufacturing, 34–35
  production control, 59

Kanban system, 33, 39–42, 50, 54
  CONWIP emulating, 46
  features of, 41
  multiple product, 44
  single-function, 44
"Keep the Product Moving," 53
Kelton, W. David, 68
Kingman equation, 68, 69
Kipling, Rudyard, 73, 76, 77, 100
Kitting, 59, 70
Koppel, Lowell B., 128

Labor
  cheap, 111–14
  hourly, 15
  subdivision of, 75, 76, 103
Lagrange multiplier optimization method, 139
Lasers, 89
Law, Avrill M., 68
"The Laxian Key" (Sheckley), 13–14
Lean manufacturing, 8, 98
  true understanding of, 107
Leather belting, 110
LeMay, Curtis, 82, 116
"Less than or equal," 20
Light bulb joke, 118
Linear programming, 19–20
  market constraints and, 27
  in Microsoft Excel, 135–41
Line-of-balance problem (LOB), 50
The Little Prince (Saint Exupéry), 82
Little's Law, 7, 68
LOB. *See* Line-of-balance problem
Longbow, 112

INDEX

Franklin on, 113, 114
Napoleon on, 113
rate of fire of, 114
Lot delay, 85
Lot splitting, 84
encouraging, 97

Make-to-forecast, 130
Make-to-order, 130
Mare Island Navy Yard, 41
Market constraints, 27–28
linear programming and, 27
Massachusetts Institute of Technology (MIT), 128
Matchsticks-and-dice simulation, ix, x, 62–67, 71
after 823 die rolls, 64
after 2,149 die rolls, 64
average die roll in, 62, 63
average production rate of, 65
beginning setup for, 62
first die roll for, 63
long-term steady status of, 65
performance graph of, 65
"rolling a six every time" in, 77, 120
second set of die rolls in, 64
Material transfer time, 98
Matrix row operations, 23
Melville, Herman, 38
Mercury vapor lamp, 115
ultraviolet radiation from, 116
Microsoft Excel, 26
answer report in, 139
cell formulas in, 136
changing cells in, 136
constraints in, 136, 138
linear programming in, 135–41
Product Mix sheet in, 136
sensitivity report in, 139, 140
Solver interface in, 137
Solver reports in, 138

spreadsheet setup in, 135
target cell in, 136
Miniè, Claude, 94
MIT. *See* Massachusetts Institute of Technology
Mixed-model production, 91
Modern Times, 73
Money
good money after bad, 13
time is, 2
"The Monkey's Paw" (Jacobs), 6
Motion efficiency, 92, 118–20
Muda. *See* Waste
Mueller Machine Tool Company, 98
Music, 38, 54
My Life and Work (Ford), 10, 73

Napoleon, Louis, 113
National Semiconductor, 127
Norwood, Edwin P., 127

Offshoring, 60
Ohno, Taiichi, 12, 82
on sunk costs, 12
Omark, 90
On War (Clausewitz), 107
Operating expense, 7, 30
Operation, 78
Opportunity cost, 3–5, 29, 30
Overhead absorption, 7, 16
Overhead conveyors, 120
Overtime, 28–29

Paced lines, 50, 54
Pareto chart, 81
Performance measurements, 6, 15
dysfunctional, 8–11
objective, 7
Peters, Tom, 108
PFR. *See* Plug flow reactor
Pig-swallowing, 33–34

INDEX

Pivot column, 22
Pivot element, 22, 23
Pivot row, 23
Plug flow reactor (PFR), 50–52, 84, 101
Pneumatic conveyor, 52
Pneumatic tubes, 53
Poisson distribution, 69
Poka-yoke. *See* Error-proofing
Poor Richard's Almanack (Franklin), 3
Preventive maintenance, 126–27, 133
   Taylor on, 127
Price concession, 28
Pricing policies, 130
Process capability, 57
   index, 87
   study, 88
Process control, automatic, 88
Process delay, 85
Process flowcharting, 77
Processing, 79, 100
Product marginal profits, 21
Production family, 100
Production Order Quantity Model, 90
Profit
   differential, 29
   marginal, 30
Pull-type production, 1, 34
   summary of, 54
Purchasing
   cheap, 114–15
   dysfunctional, 9
Push production system, 33
   responsiveness and, 34
Push-pull interface, 11

Quality control. *See* In-line quality control
Quality problems, 61–62
Quick-clamping flanges, 95, 96

Radio frequency identification (RFID), 80

"Receive material," 79
Redburn: His First Voyage (Melville), 38
Regulations for the Order and Discipline of the Troops of the United States (Steuben), 37
Reliability problems, 61–62
Render, Barry, 25
Return on investment (ROI), 7, 14–15
   cheap equipment and, 111
Revenue, 15, 16
   differential, 29
   marginal, 30
RFID. *See* Radio frequency identification
Rhythmic timing, 37
Rifles
   Baker Rifle, 93
   evolution of, 94
   flintlock era of, 93
   improvements in, 94
   loading drills for, 92
   measuring powder charge for, 93
   Tower Musket (Brown Bess), 93, 94, 114
   von Dreyse, 94
Rigg, James L., 4
RMS. *See* Root-mean square
Robinson, Alan, 8
ROI. *See* Return on investment
Rolled throughput yield (RTY), 121
Root-mean square (RMS), 125
RTY. *See* Rolled throughput yield

Safety cushion, 66
"Safety stocks," x, 44
   product wheel, 43
Saint Exupéry, Antoine de, 82
Saint Nazaire, France, 116–17
Scaffold, nonstooping, 118
Schonberger, Richard J., 51
   on friction, 108
Scrap, 120

INDEX

rate, 121
Scrooge, Ebenezer, 115
Securities and Exchange Commission, 11
Self-check systems
   at Ford Motor Company, 61
   in-line quality control as, 122–24, 126, 132
   Shingo on, 122, 123
Self-through data, 129
"Sending good money after bad," 13
Sequence inspections, 132
Serial batch, 84, 85
Shadow prices, 20, 25, 139, 140, 141
Sharpe's Rifles, 93
Sheckley, Robert, 13–14
Shingo, Shigeo, 91
   on self-check systems, 122, 123
   on stem tighteners, 123
Simplex method, 20–26
Simplex tableau, 21
   initial, 22
Simplified market pull (SMP), 45–46, 54
Single-minute exchange of die (SMED), 19, 84, 90–97
   external setup of, 91
   internal setup of, 91
   key concepts behind, 97
   military origins of, 92–95
Single-server queue, 67–70
   cycle time of, 67
Single-unit flow, 34, 89
Single-unit heat treatment equipment, 90
Single-unit processing, 84–90, 91
Skillets, 51, 52, 77
   drawback, 77
Slack variable, 21, 26
Sloan, Alfred, 8
SMED. *See* Single-minute exchange of die

Smith, Wayne, 43
SMP. *See* Simplified market pull
Sorenson, Charles, 36
Source inspections
   horizontal, 124
   in-line quality control as, 124–25, 126, 132
   vertical, 124
Spaghetti diagram, 97–98
SPC. *See* Statistical process control (SPC)
Spearman, Mark, 11
Speed limits, 117
Split thread bolts, 95, 96
Standard, Charles, 8, 9
   on pig-swallowing, 33–34
Standard deviation, 75
Statistical process control (SPC), 47, 87
Statistical throughput chart, 48
Statistical throughput control (STC), 47–49
   simulations, 49
STC. *See* Statistical throughput control
Steady-state rate, 67
Stem tighteners, 123
Steuben, Baron von, 37
Storage, 79
Successive check systems, 126
Suppliers, Inputs, Processes, Outputs, and Customers Model, 58
Supply-chain, 60
   inefficiencies in, 133
   instability in, 129–31
   logistics, 131–32
   management, 60, 128–31
Supply-chain dependence, 59
Surplus resource, 24
Suvorov, Aleksandr v., 2
Synchronized production, 35–38
System Company, 100

Taguchi robust design problem, 88

INDEX

Takt time, 35–38, 54
Taylor, Frederick Winslow, 110, 118
   on preventive maintenance, 127
Theory of Constraints (TOC), x, 1
   variable costs in, 15–17
Therbligs, 79, 80
Threshold parameter, 69, 70
Throughput, 7, 8, 15, 30
Time accounting, 80, 103
"Time is money," 2
TOC. See Theory of Constraints
Total-productive maintenance (TPM), 127
Tower Musket (Brown Bess), 93, 114
Toyoda, Sakichi, 125
Toyota
   advertisement, 120
   heijunka concept at, ix, x
   waste and, 109
TPM. See Total-productive maintenance
Traffic jams, 66
Transform, 80, 103
Transportation, 78, 79, 80, 103
Truck sharing, 129
Turbide, Dave, 45
Turret lathe, 100
Two Years Before the Mast (Dana), 37–38

Unitary machines, 97, 100–103, 104
   first civilian, 102
   Gatling gun as, 101
Unstable feedback loop, 128
U.S. News & World Report, 120

Value stream analysis, 77, 100, 103
Variability-utilization-time equation (VUT), 68
Variance, 75
Variation
   batch-to-batch, 87
   common cause, x

   effects of, 62–67
   in environment, 58
   increasing, 85
   in machines, 57
   in manpower, 57
   in materials, 58
   in measurements, 58
   in methods, 58
   random, x
   reduction of, 103–4
   sources of, 57–62
   summary of, 70
Vendor-managed inventory (VMI), 129
Visual control system, 40
VMI. See Vendor-managed inventory
Von Dreyse rifle, 94
VUT. See Variability-utilization-time equation

Wages, 112
"Waiting to match," 59, 70
Walker, Bill, 129
Waste, 91, 107, 122
   Bennett on, 108–9
   Ford on, 108–9, 132
   Toyota and, 109
The Way to Wealth (Franklin), 3, 9
Wellington, Duke of, 112
Wheelwright, Steven C., 36
White, Maunsel, 110
Willow Run bomber plant, 36
WIP. See Work-in-process
Woolley, Edward Mott, 118
Work cells, 97, 98–100
   characteristics of, 99
Work holders, rigid, 120
Workflow analysis, 78–80
Work-in-process (WIP), 4, 67
Workpiece, 51
World War II, 116

ABOUT THE AUTHOR

William A. Levinson, principal of Levinson Productivity Systems, P.C., has authored and co-authored several books with the American Society for Quality including *Lean Enterprise: A Synergistic Approach to Minimizing Waste, ISO 9000 at the Front Line, SPC Essentials and Productivity Improvement, Leading the Way to Competitive Excellence,* and *The Way of Strategy.*

He is also the author of *Henry Ford's Lean Vision: Enduring Principles from the First Ford Motor Plant* (Productivity Press, 2002).

TS 155 .L3668 2007
Levinson, William A., 1957-
Beyond the theory of
 constraints

MAY 1 2 2008